At Home in Space

Title Page and Dustjacket Cover Photos: NASA—Scene of Phase III, Human Tended International Space Station produced by John Frassanito and Associates.

Unless otherwise indicated, all photos throughout At Home in Space *receive a NASA credit line.*

At Home in Space

by

Howard Benedict

Edited by
Myrtle D. Malone

Published by
Pioneer Publications, Inc.
Bill Levisohn, Publisher

Pioneer Publications, Inc.
12345 Jones Road, Suite 103
Houston, Texas 77070-4843

International Standard Book Number: 1-881547-19-1
Library of Congress Catalog Card Number 95:67719
Pioneer Publications, Inc., Houston, Texas 77070-4843

Printed in Thailand.

Acknowledgment

Pioneer Publications, Inc. gratefully acknowledges the assistance provided by the National Aeronautics and Space Administration in the publication of this volume, particularly in providing photographs used throughout *At Home in Space*, as well as many of the photographs used by various companies to illustrate their participation in the space program. Our special thanks to Mike Gentry, Chief of the Still Photo Lab at Johnson Space Center, for his photographic help. Unless otherwise indicated, photographs in *At Home in Space* receive a NASA credit line and are not subject to copyright.

Contents

At Home in Space

1 Introduction
3 Chapter One — The Assault on Space Has Begun
8 Chapter Two — For the Benefit of All Mankind
13 Chapter Three — Now Let the Other Countries Try to Catch Us
17 Chapter Four — Man is the Deciding Element
25 Chapter Five — A Man is Floating Free in Space
28 Chapter Six — We're Tumbling End Over End
32 Chapter Seven — I've Got a Fire in the Cockpit!
37 Chapter Eight — A Vast, Lonely, Forbidding Place
43 Chapter Nine — One Giant Leap for Mankind
51 Chapter Ten — Houston, We've Had a Problem
58 Chapter Eleven — Man Must Explore
64 Chapter Twelve — It's Been a Good Home
69 Chapter Thirteen — Soyuz and Apollo Are Shaking Hands
73 Chapter Fourteen — A Fabulous Flying Machine
82 Chapter Fifteen — A Major Malfunction
88 Chapter Sixteen — Piece of Cake
93 Chapter Seventeen — A Force for Peace and Progress
99 Chapter Eighteen — The Economic Boon Will Be Enormous

Company Editorials

106 Rockwell
110 Instron Corporation
114 Hughes Electronics Corporation
118 Martin Marietta Corporation
122 Motorola
126 Network Systems Corporation
132 DATATAPE Incorporated
134 Johnson Engineering Corporation
136 AlliedSignal Aerospace
138 Space Vector Corporation
140 Aerotherm Corporation
142 ManTech International Corporation
144 Booz·Allen & Hamilton
146 SPACEHAB
148 Evans & Sutherland
150 Barry Controls
152 Harris Corporation
154 McDonnell Douglas Aerospace
156 Applied Solar Energy Corporation
159 Digital
160 Hughes Training, Inc.– Link Division
162 USBI Co.
163 Barrios Technology
164 Silicon Graphics, Inc.
165 Eaton Corporation
168 Index

Introduction

A quarter century ago, a skilled, competent National Aeronautics and Space Administration team successfully dispatched the first human beings to another world and a sensational landing on the moon — and history was made.

NASA, brimming with optimism, had hoped the footprints left there in 1969 by Neil Armstrong and Buzz Aldrin would catapult it into an even more glorious future in space.

Indeed, the U.S. had won the race to the moon, the Soviets left far behind, in a high-stakes contest for national prestige. The goal achieved, the impetus was gone. NASA had hyped Apollo as primarily a competition with the Russians. It had no long-term strategy and it did not follow up. A long, troubled period began for the agency. Budgets were slashed and tens of thousands of workers lost their jobs.

NASA administrator Thomas Paine tried to turn the tide in 1970. To a presidential panel reviewing post-Apollo space policy, he proposed permanent manned space stations, a research base on the moon, manned flights to Mars before 1990 and a fleet of reflyable space planes to shuttle between Earth and orbit. He envisioned eventually transporting hundreds of humans to build homes in space, to establish colonies throughout the solar system.

But Paine and other officials had misread the national mood. The political climate of the time was not conducive to such grandiose ideas. The country was torn by Vietnam. The racial and campus unrest of the time had prompted much of the press and many politicians to view the space program as a symbol of "misplaced national priorities." With détente, the Soviet threat no longer seemed so obvious.

NASA's dreams were derailed by Congress and the administration of President Richard M. Nixon. Paine was told to scrap his plans for large space stations, a moon base and manned trips to Mars. The only major man-in-space proposal to survive was the space plane, and its projected funding was cut deeply.

The nation then built the most sophisticated flying machine that mankind has ever seen — the Space Shuttle. Except for one heart-wrenching tragedy, the explosion of *Challenger* in 1986, the shuttle has flown successfully into orbit scores of times. But, unlike the Apollo project — in which the U. S. had a space program with a defined goal — each shuttle flight is an event in itself. Crews deploy a satellite, repair or retrieve a satellite, or conduct experiments. One flight is rarely tied to the next, or to the future, as the Mercury, Gemini and Apollo flights were. So, one cannot look at a series of shuttle flights as a mission, or as a true space program, because there is no ultimate destination.

Left: A footprint left on the moon by Apollo 11 astronauts. Having won the race to the moon, NASA officials envisioned permanent manned space stations, a research base on the moon, manned flights to Mars, human colonies throughout the solar system, and a fleet of reflyable space planes to shuttle between Earth and orbit. The only man-in-space proposal to survive space program funding cuts was the space shuttle.

In 1986, Aaron Cohen, director of the Johnson Space Center, exhibited this model of the proposed space station.

Until the United States puts together a program that utilizes the shuttle as a true transportation system, each and every flight will remain an event.

Now, after several years of setbacks, indecision and stretched budgets, NASA is establishing such a program for the shuttle. That mission is an international space station — a permanent outpost to be built with foreign partners from Europe, Canada, Japan and old rival Russia. After years of missteps and delays, this multi-nation cooperative effort appears to be on track toward starting construction of the station in about two years, using both the shuttle and Soviet and European rockets to ferry up men and building materials.

The effort could serve as a blueprint for embarking to even more exotic destinations in space. By uniting talent, technology and money, astronauts from many lands could set sail on voyages to establish research bases on the moon, Mars and elsewhere. Then, perhaps, in the 21st century, NASA can achieve its Apollo-era dream of those homes in space — colonies of hundreds working and thriving beyond Earth's atmosphere.

Even most of NASA's most severe critics hope the agency, through an international effort, can turn it around and once again rekindle our imagination.

As a long-time aerospace writer for The Associated Press and now as executive director of the Mercury Seven Foundation, I have had the privilege to witness space history from the very beginning. In this book I relate that history — from the early days of exploding rockets, to the first tentative man-in-space steps in Mercury and Gemini, to the glory days of Apollo, to the Space Shuttle era, to the uncertainties of today — and which direction the world may be headed in space.

— Howard Benedict

They called the rocket "Bumper." It was a makeshift vehicle of two stages, built by combining a German V-2 and a U.S. Army WAC-Corporal. The launching pad was a plumber's nightmare — an 80-foot spider web of pipe and scaffolding festooned with cables and wires. Control centers included tents, sandbagged trenches and an old Sherman tank. A Jeep raced back and forth on a sandy road leading to the launch site to frighten away alligators and poisonous snakes. After five days of frustrating delays, battling the heat and mosquitoes, the 30-man launch team pushed the button and Bumper blasted off at 9:29 a.m. July 24, 1950. The rocket sped to a height of 10 miles and dropped into the Atlantic Ocean 50 miles off the Florida coast. That was the first rocket launched from Cape Canaveral, the new test center for America's military missiles. Other cruise missiles followed: the Lark, Matador, Snark and Bomarc. Then came ballistic missiles: Redstone, Thor, Jupiter, Polaris, Atlas and Titan, with ever-increasing range. As the ballistic missiles became operational, space visionaries suggested using them to boost satellites into orbit around the Earth to study space. Among them was Wernher von Braun, the German rocket scientist who helped develop the V-2 and who, after the war, brought a team of German rocketeers to the United States to work on the army missile program. That team was instrumental in developing America's first ballistic missile, the Redstone, and in 1956, von Braun proposed propelling a scientific satellite into orbit by attaching an upper stage to the Redstone. Finally, on January 31, 1958, a Redstone with an upper stage hurled Explorer I into space. (U.S. Air Force photo.)

Chapter One

'The Assault on Space Has Begun'

The evening was cool, the conversation crisp as the Army's top brass enjoyed before-dinner cocktails in the officer's mess at the Redstone Arsenal in Huntsville, Alabama. The occasion was an orientation tour for Neil H. McElroy, recently nominated by President Eisenhower to replace Charles E. Wilson as secretary of defense. The guests included Secretary of the Army Wilber M. Brucker; Gen. Lyman L. Lemnitzer, chief of staff; Maj. Gen. John B. Medaris, commander of the Army Ballistic Missile Agency at Redstone, and Wernher von Braun, the brilliant former German rocketeer whose team had developed the Redstone and Jupiter missiles for the Army.

McElroy was there primarily to learn about the Redstone, a 200-mile-range weapon, and the 1,500-mile Jupiter, America's most advanced missile. But Medaris and von Braun used his visit to plead their oft-rejected case to launch a satellite into orbit about the Earth with a souped-up Redstone. Throughout the day and into the cocktail hour, they pressed their argument with charts and figures and words. They contended that Project Vanguard, America's satellite effort based on an untried rocket, was doomed to fail and that the Soviet Union would embarrass the United States by orbiting the first satellite.

Halfway through the party, their arguments became academic. Gordon Harris, Medaris' public information officer, rushed up to McElroy, von Braun and Medaris and blurted out. "It has just been announced over the radio that the Russians have put up a successful satellite."

The launching, at 10:30 p.m. Moscow time, October 4, 1957, marked the birth of the space age. Few eras in history have had such precise beginnings.

Harris' announcement was followed by a moment of stunned silence. Then von Braun erupted on McElroy: "We knew they were going to do it," he said excitedly. "Vanguard will never make it. We have the hardware on the shelf. For God's sake, turn us loose and let us do something. We can put up a satellite in 60 days, Mr. McElroy. Just give us the green light and 60 days."

McElroy could make no immediate commitment because be had not yet been confirmed as defense secretary by the Senate. But, as von Braun said, "Mr. McElroy had a lot to think about when he went back to Washington." The first thing the secretary-designate and the rest of government encountered was the electrifying impact of the satellite that the Soviets called Sputnik, which meant traveling companion.

The 184-pound satellite, circling the globe as far out as 569 miles every 96 minutes at more than 17,000 mph, shocked the world, especially the

Dr. Wernher von Braun.

Americans, who were embarrassed and apprehensive.

People around the globe stood in the twilight or early dawn, their eyes fixed on the heavens, searching for an elusive pinpoint of light moving across the sky. Amateur radio buffs tuned in on the incessant beep-beep-beep coming from space. They were fascinated and frightened, knowing man had put it there and wondering where it could possibly lead.

Until then, the United States had been considered the world's unchallenged technological leader, with the Soviets trailing far behind. Sputnik not only symbolized the emergence of the Soviet Union as a technologically-advanced society, but demonstrated for the first time that the Soviet military had the rocket power to deliver nuclear weapons across continents and oceans to the United States. America's most powerful rocket could travel only 1,500 miles.

American skies had been violated. During World War II, no Nazi or Japanese aircraft had penetrated U.S. air space, but here was Sputnik, made-in-Russia, passing overhead four or five times a day. The

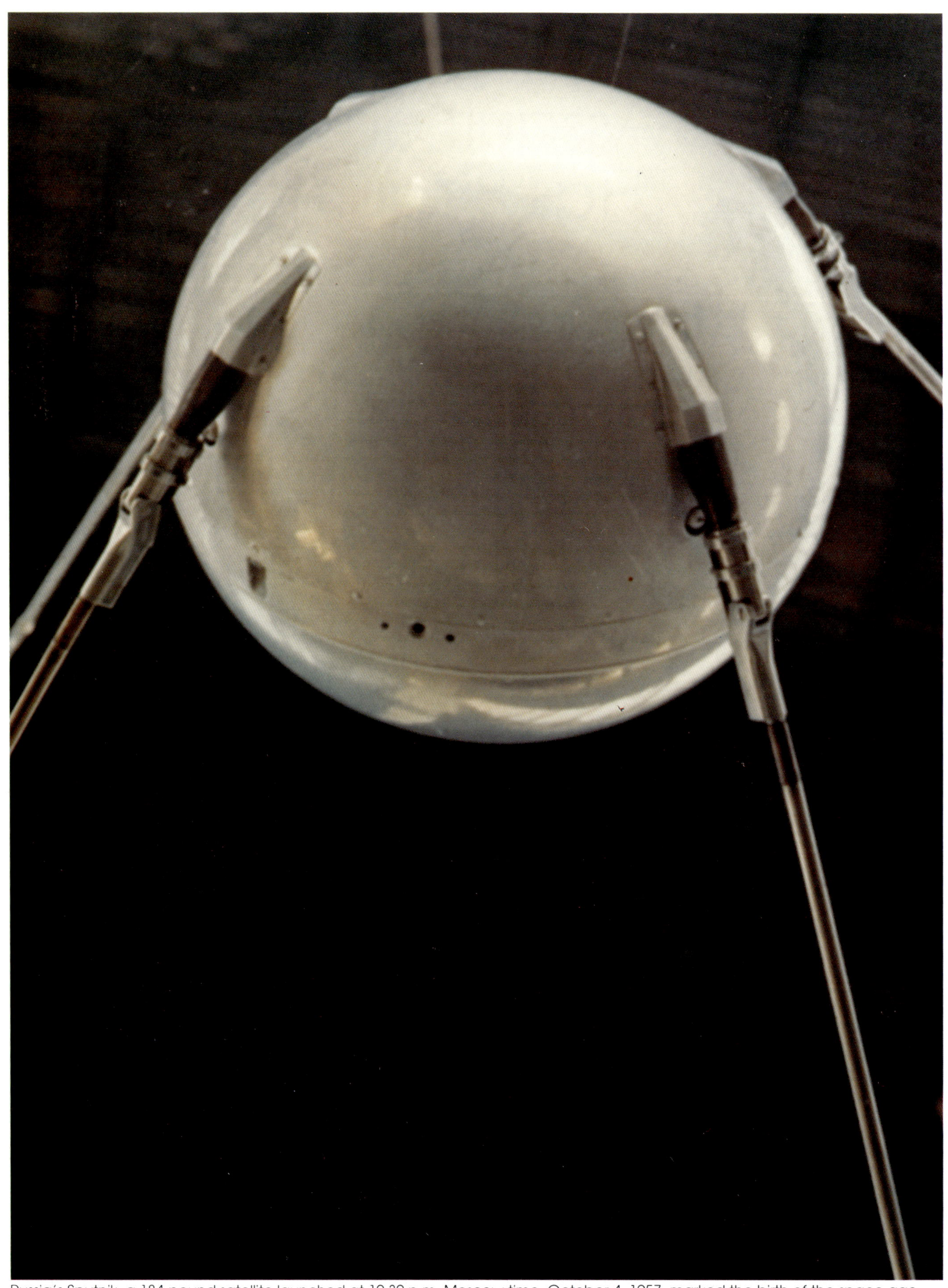

Russia's Sputnik, a 184-pound satellite launched at 10:30 p.m. Moscow time, October 4, 1957, marked the birth of the space age.

Eisenhower administration was caught by surprise. In general, its spokesmen tried to downplay the significance, writing it off as a "silly bauble," a "neat technical trick," and as "one shot in an outer space basketball game." Eisenhower told a news conference, "After all, the Russians have only put one small ball in the air" and it did not threaten national security.

Soviet Premier Nikita Khrushchev saw it differently, boasting that it demonstrated the superiority of communism over capitalism. "People of the world are pointing to the satellite," he stated. "They are saying the U.S. has been beaten."

Khrushchev long had been aware of the psychological value of orbiting the first satellite, and he had supported Sergey Korolov, the nation's chief rocket engineer, when he proposed boosting a payload into orbit with the new and powerful R-7 rocket. Others in the Kremlin had belittled the suggestion as a "game," but Khrushchev argued it would bring prestige to the Soviet Union and demonstrate unequivocally the existence of a long-range missile system capable of striking the United States.

Korolov, a victim of the Stalinist purges of the intelligentsia in the 1930s and a survivor of the infamous Kolyma Camp of the Gulag Archipelago, and Khrushchev communicated frequently as the launch of Sputnik 1 neared from a rocket base north of the central Soviet town of Tyuratam in Kazakhstan. The chief engineer literally lived day and night with the project, operating from a small wood-frame house that was a 15-minute walk from the rocket assembly building in one direction and 10 minutes from the launch pad in another.

Sputnik was a simple test satellite put together in a machine shop. It was a basketball-size, 184-pound ball of aluminum alloys with four spring-loaded whip antennas. Inside were two continuously signaling transmitters powered by chemical batteries. It was fitted atop the rocket and a pointed metal nose cone was slipped over it.

On launch day, Korolov took his place at a wooden desk, microphone in hand, in a steel-walled room built into the concrete launch pad, about 300 feet from the R-7 rocket. After many exasperating technical delays, the countdown finally reached zero and the rocket was off with a roar, lighting the night sky. Both rocket stages fired perfectly and thrust the satellite into orbit. Gravity still tugged the free-flying payload back toward Earth on a descending arc, but so great was its speed that the horizon of the Earth receded from the falling object at the same rate and Sputnik remained locked in orbit.

Ninety minutes later, when the world's first artificial satellite passed over the launch pad, its beep-beep-beep was played over loudspeakers. The rocket workers cheered, and Korolov, who for years had been a strong believer in space exploration, told them, "Today the dreams of the best sons of mankind have come true. The assault on space has begun."

Two days later, the Soviets announced they had exploded a "powerful hydrogen device of a new design," and they began rattling their rockets belligerently. Members of the North Atlantic Treaty Organization received letters from Khrushchev threatening them with H-bomb destruction if they permitted U.S. missiles to be based on their territory. And the United States was threatened with rocket retaliation if it interfered directly in Lebanon's struggle with the Moscow-backed United Arab Republic. The cold war was heating up.

Sputnik brought cries of shock and alarm from American politicians, scientists and other figures. Edward Teller, father of the atomic bomb and then associate director of the Livermore Radiation Laboratory, warned the future might now belong to the Soviet Union and proclaimed the United States had "lost a battle more important and greater than Pearl Harbor." Dr. James R. Killian Jr., the president of Massachusetts Institute of Technology, who in the wake of Sputnik became the first White House science advisor, recalled later that Sputnik "created a crisis of confidence that swept this country like a windblown forest fire."

No one was more disturbed or frustrated by the Russian feat than Wernher von Braun, who had pleaded vainly for more than three years for permission to launch an American satellite by adding upper stages to his proven Redstone ballistic missile. Why was he so confident? When the Army commissioned him and his rocketeers to develop the Jupiter missile, they first had to find out if a nose cone fired over a 1,500-mile-range could survive the severe re-entry heat as it dived on target with its nuclear payload. Von Braun's team came up with the idea of having the smaller Redstone hurl a cluster of 14 small solid-fuel Sergeant rockets away from the nation's ballistic missile test range at Cape Canaveral on the Florida east coast. The cluster — 11 rockets in a second stage, three in a third — would separate to propel a small dummy warhead far out over the Atlantic at the same re-entry speed as the Jupiter payload.

While von Braun was preparing to fire the first Jupiter-C, called Missile 27, from the Cape in September 1956, the Pentagon became suspicious that he was planning to hurl a satellite into orbit and apologize later. General Medaris was ordered to personally inspect the rocket to make certain it did not have a live fourth stage. It didn't, and the Jupiter-C performed flawlessly, hurling a dummy warhead 600 miles high and 3,300 miles down range at 16,000 miles an hour — farther and faster than any U.S. missile had flown at that time.

"We knew then we could put a satellite in orbit," von Braun said. "All we had to do was to add a single solid rocket as a fourth stage." The Army and von Braun asked for permission to try. A panel of government scientists decided that any American satellite project should be launched with a rocket that did not have military origins and recommended development of a new booster called Vanguard as a contribution to International Geophysical Year, during which scientists around the globe were to make a concentrated study of the Earth between July 1957 and December 1958. They said a rocket with a peaceful background would have "more dignity" for a scientific project like IGY. President Eisenhower agreed.

Snorted von Braun: "I'm all for dignity, but this is a cold war tool. How dignified would our position really be if a man-made star of unknown origin suddenly appeared in our skies?"

Despite his pleas, von Braun was told to store his

backup missile, No. 29, in a large shed where it gathered dust for months — to be unleashed only after the Soviets pulled their second shocker.

On November 3, less than a month after the launching of the first Sputnik, Sputnik 2 shot into orbit. It weighed an astounding 1,130 pounds and had a passenger, a dog named Laika. A life support system kept the animal alive for more than 100 hours, and, although the dog died, it was a clear indication the Russians were thinking about sending humans into space. Sputnik 2's heavy weight was a complete surprise. The Vanguard was being groomed to carry a satellite weighing only about three pounds, while the Jupiter-C at best could hoist about 35 pounds.

Why were the Soviets able to orbit such large satellites? The answer lay in American leadership in developing small, sophisticated electronic components for nuclear warheads, enabling the U.S. to build smaller rockets to deliver these weapons. The Russians, without this sophistication, had to construct large boosters from the start. Thus, they had the bigger weight-lifting capability when man realized he could use this power to explore space.

Khrushchev, addressing the Supreme Soviet, declared: "Our Sputniks are circling the world. . . Wars are not needed for the victory of socialism. Imperialists will not be able to stop the forward march of society to communism, whatever their efforts."

There was more panic in America. The Vanguard team was told to accelerate its effort, and von Braun was directed to dust off Missile 29 and prepare it for launching a satellite should the Vanguard fail a December attempt.

Technicians prepare Explorer I, the United States' first artificial Earth satellite, for launch at Cape Canaveral, Florida. Shaped like a stovepipe, the satellite weighed just under 31 pounds.

At Cape Canaveral, the Vanguard team strained to met the December deadline, overcoming one technical problem after another and finally setting a December 4 launch date. As the day approached, thousands of people and reporters from around the world poured into the Cape area to see if the Americans could begin meeting the Soviet challenge.

Weather and mechanical glitches caused several delays, but finally, at 11:45 a.m. on December 6, the Vanguard's first stage engine spewed brilliant white flame. Two seconds later, someone in the blockhouse screamed, "Look out! Oh, God, no!" Another voice shouted, "Duck!" To propulsion engineer Kurt Stehling it seemed "as if the gates of hell had opened up." With the world watching, the Vanguard rocket rose about four feet off its launch pad, shuttered slightly and collapsed on itself, exploding into a huge fireball. The tiny satellite, weighing only 3.2 pounds, was flung free of the wreckage and lay in nearby underbrush, its radio transmitter beeping, a forlorn symbol of a massive American failure. Investigators said low fuel tank pressure was the apparent cause.

Khrushchev derisively called the Vanguard satellite an "orange." Headlines around the globe proclaimed: "Stayputnik," "Ike's Sputnik is a Flopnik," "Oh, What a Flopnik."

It was a very bleak day for America. To its citizens, the failure generated feelings of lost confidence, wounded pride, confusion — and awe at the Soviet achievements. What is wrong, they asked. How could these supposedly inferior, baggy-pants communists develop more rocket power than the United States? Have we gone too soft? Have we, in the great rush to the suburbs, wasted our industrial and technological power on producing new model automobiles and gadgets? What's wrong with our schools? Among the American weaknesses that came under fire was the insufficient support for science and engineering and the neglected state of the education system. The U.S. was graduating 22,000 engineers a year, the Soviets, 66,000. The Eisenhower administration was criticized for its complacency, for allowing the Soviets to get so far ahead in rocketry and space flight. In Congress, there were shouts for action, for reappraisal, to end the humiliation.

Against this backdrop the Army and its mainly-German rocketmen transported the Jupiter-C — renamed a Juno 1, in a weak attempt to disguise its military background — to Cape Canaveral in January 1958. Von Braun reserved a four-day period starting January 29. As with Vanguard, there were weather and technical problems, but they were overcome, and at 10:48 p.m. January 31, the rocket ignited and etched a fiery path in the sky as it blazed out over the Atlantic Ocean.

Although early tracking indicated the booster had performed well, definite confirmation of orbit would have to come from tracking stations in California after the satellite had made one nearly complete revolution of the globe. When that confirmation came, there was near bedlam of back-slapping, hugging and

The Explorer I launch, January 31, 1958, catapulted America into the space age. A Geiger counter aboard, developed by physicist James A. Van Allen, discovered that the Earth is surrounded by huge bands of high-energy radiation, later named after him.

congratulating at Cape Canaveral and elsewhere in the country.

"It was one of the great moments of my life," said von Braun. "I only regret we didn't do it earlier."

Across the land, Americans breathed a collective sign of relief. Von Braun became a national hero, featured on magazine covers and honored at the White House, where President Eisenhower, who had supported the civilian Vanguard over the military Redstone, presented him with the Distinguished Federal Civilian Service Award.

The satellite was named Explorer 1. It weighed only 31 pounds, but a Geiger counter aboard made the first discovery of the new era — that the Earth is surrounded by huge bands of high-energy radiation composed of particles trapped in our planet's magnetic field. The Geiger counter was developed by James A. Van Allen, a State University of Iowa physicist. Scientists honored Van Allen by naming the belts after him.

With Explorer 1, America catapulted into the space age and a fierce competition with the Soviet Union over who would dominate this new ocean.

Chapter Two

'For the Benefit of All Mankind'

President Eisenhower had a problem in early 1958. The nation, still stung by Sputnik but buoyed by Explorer, was demanding that the president take whatever action was needed to catch the Soviets in space and missilery. Both houses of Congress approved large sums of money for a number of response programs; both established space committees and urged the administration to set up a Department of Space or a Department of Science. The world accepted as fact that the superpowers were in a high-stakes race, but Eisenhower did not want to acknowledge such a contest. He did not want to invest in an expensive space race, but with pressure mounting, he had to do something.

The president sought advice from his new science advisor, James R. Killian Jr. In an Oval Office meeting, Killian told Eisenhower that, considering the mood of the country, he had little choice but to accept many of the congressional proposals. He reluctantly agreed, and things began to happen. An arbitrary upper limit on the defense budget was removed, and military missile programs were given what amounted to blank checks. Most forms of government-subsidized research and development began to grow. A flood of federal dollars overhauled the nation's education system at all levels. Loans provided money for more than 1.5 million college students in a decade, producing 15,000 Ph.D.'s. The education initiatives produced an unmatched generation of scientists and engineers who became part of the U.S. response to the Soviets.

Killian also proposed that a space agency, not a full department, be created and that it be nonmilitary. About the same time, Texas Sen. Lyndon B. Johnson presented the findings of his Preparedness Investigation Committee calling for establishment of an independent civilian space agency. Who would manage this new effort? Killian focused on the National Advisory Committee for Aeronautics, NACA, a little-known group of research facilities which provided other government bodies like the Air Force and Navy with the fruits of its work. NACA possessed some of the best engineering talent in the country, was under civilian control, which Eisenhower preferred, was too little known to have been caught up in partisan politics, and was a stranger to the red tape of government bureaucracy. This relatively obscure orga-

Eight of the twelve members of the National Advisory Committee for Aeronautics are pictured in this photo taken at Langley Field, Virginia, May 23, 1934, on the occasion of the 9th Annual Aircraft Engineering Research Conference sponsored under the auspices of NACA. Seated, left to right, are: Brig. Gen. Charles A. Lindberg, USAFR; Vice Admiral Arthur B. Cook, USN; Charles G. Abbot, Secretary of the Smithsonian Institution; Dr. Joseph S. Ames, Committee Chairman; Orville Wright; Edward P. Warner; Fleet Admiral Ernest J. King, USN; and Eugene L. Vidal, Director, Bureau of Air Commerce. Standing, left to right, are: Dr. George W. Lewis, Director, NACA; Dr. Henry J. E. Reid, Director, Langley Aeronautical Laboratory; and John F. Victory, Executive Secretary, NACA.

nization was handed the job of challenging and overcoming the Soviet space lead.

Eisenhower in April 1958 sent Congress a bill to create a civilian space agency. Under the guidance of Senator Johnson, legislators established the National Aeronautics and Space Administration to direct all nonmilitary space projects and a National Aeronautics and Space Council to advise the president. NACA's five laboratories and 8,000 engineers and technicians were drawn under the new NASA umbrella, as were the Vanguard satellite project and the Jet Propulsion Laboratory, previously operated under an Army contract by the California Institute of Technology. While the Army reluctantly yielded control of JPL, it steadfastly refused to let NASA have the Wernher von Braun rocket team at Huntsville. "We were in the atomic age and the Army absolutely did not want to lose its major rocket development capability at a time when the Navy and Air Force were pushing strong missile projects," von Braun said.

NASA formally began business on October 1, 1958, with a mandate to use aeronautical and space capabilities "for the benefit of all mankind." The program was to be conducted in the open, nothing withheld from the public, in contrast with the secretiveness of Soviet space activities. Eisenhower named T. Keith Glennan, president of the Case Institute of Technology and a former member of the Atomic Energy Commission, as the first administrator of the new agency. Hugh L. Dryden, who headed NACA, was named his deputy. It started with a budget of $330 million.

NASA's role was to enter space for scientific, peaceful purposes. There was an unwritten goal: Put America in front in space and keep it there. Glennan and Dryden went about doing that. But first, they needed a plan. While the immediate goal was to develop and launch satellites that could do useful things for mankind — such as forecasting the weather and improving communications — agency officials knew from the beginning that the effort must include a manned space program. The Russians indicated they were headed in that direction, and America could do no less.

Yet, Eisenhower mentioned no man-in-space assignment for NASA in constituting the agency. Into this vacuum stepped the Army, the Air Force and the Navy, each with its own plan to loft humans into space. But the president and his advisers were convinced there was no valid role for the military in space and they killed the services' proposals. Under Killian's urging, Eisenhower assigned NASA responsibility for manned space flight, and Glennan established a Space Task Group at its Langley Research Center in Virginia to manage the effort.

The project was called Mercury, after the fleet messenger of the ancient Roman gods. It was headed by Robert R. Gilruth, a brilliant engineer and manager who had directed Langley's Pilotless Aircraft Research Division.

NASA immediately called for bids from industry to build a manned spacecraft that would carry a single passenger, one that would land on water because the first 3,000 miles of flight would be over the Atlantic

After President Eisenhower assigned NASA responsibility for manned space flight, NASA's first administrator, T. Keith Glennan, established a Space Task Group at Langley Research Center in Virginia to manage the effort.

Ocean and the capsule would have to be able to float in case of an emergency shortly after liftoff. McDonnell Aircraft Co. of St. Louis won the contract to build the 1.5-ton craft, which would be nine feet tall and six feet across at its widest point.

On December 17, 1958, on the 55th anniversary of the first Wright brothers' flight, the agency said it would begin recruiting astronauts from the ranks of test pilots. They would have to be between 25 and 40 years old, in perfect physical condition, and, because of spacecraft limitations, no more than 5 feet 11 inches tall. They also would have to have a college degree. Eisenhower had decided the astronauts must come from the military services. The Army had no graduate test pilots, so the early screening list included 58 Air Force test pilots, 47 from the Navy and five from the Marines. Physically and psychologically, they probably were the most studied, examined men in history.

Seven were chosen, and on April 9, 1959, the Mercury astronauts were introduced to the public at a news conference in Washington, D.C. Their names

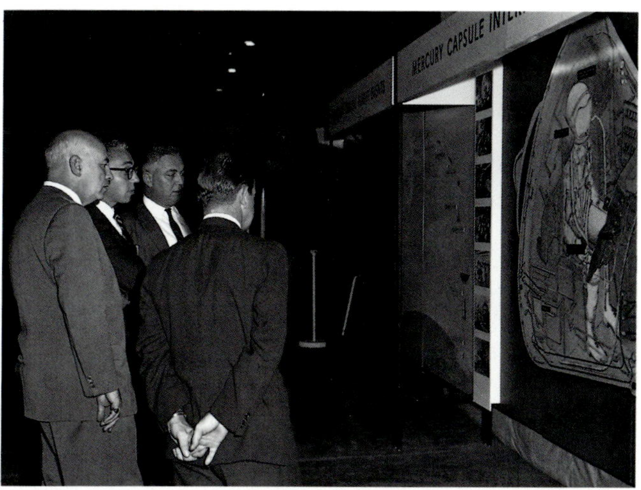

Dr. Robert R. Gilruth (left), head of Project Mercury, looks at an engineer's drawing of the Mercury capsule. Others in this 1961 picture are (left to right) D. Brainerd Holmes, Walter C. Williams and John (Shorty) Powers.

NASA's original seven astronauts: (left to right) Carpenter, Cooper, Glenn, Grissom, Schirra, Shepard, and Slayton.

and feats would be forever etched in the annals of space history. From the Navy: Lt. Comdrs. Alan B. Shepard Jr., Walter M. Schirra Jr., and Lt. Malcolm Scott Carpenter. From the Air Force: Capts. Virgil I. (Gus) Grissom, Leroy Gordon Cooper Jr., and Donald K. (Deke) Slayton. From the Marines: Col. John H. Glenn Jr.

A night view of the Manned Spacecraft Center in Houston, Texas — later named the Lyndon B. Johnson Space Center — taken in August 1965.

The astronauts, not knowing when they would fly, began their intense training at Langley Research Center, moving in 1961 to the new Manned Spacecraft Center near Houston, Texas, later renamed the Lyndon B. Johnson Space Center. These men, who were being groomed to take on the Russians, became instant heroes, hailed as the bravest of the brave for daring to volunteer to ride one of those rockets which kept blowing up at Cape Canaveral. Their pictures and biographies were plastered all over the newspapers, magazines and television screens, and their courage had yet to be tested on a rocket.

While astronauts and cosmonauts awaited development of their space chariots, the battle for supremacy in the heavens was being waged with unmanned satellites. As the decade ended, the Soviets had successfully probed the moon and orbited living creatures, including two dogs returned to Earth alive. They also launched probes to Mars and Venus, and although both fell short, it demonstrated the depth of Russia's commitment to space exploration. U.S. scientific payloads had discovered remarkable things about the Earth and the protective shell of gases around it. America also had orbited satellites for weather forecasting, communications, navigation and military surveillance.

The U.S. had launched 33 satellites, the Soviets

nine. But those nine Soviet payloads, hoisted by much larger rockets, outweighed all those of the U.S. by two-to-one. NASA chief Glennan took notice and told his advisers that if America was to win this cosmic race, it would need bigger rockets. He knew that the von Braun team at Huntsville had on the drawing board a large booster named Saturn 1, designed to generate 1.5 million pounds of thrust by clustering eight rockets powered by engines that were improved versions of those used on the Redstone and Jupiter. A single Redstone alone generated only 78,000 pounds of thrust, while the Atlas being developed for Mercury orbital flights delivered 360,000 pounds. In early 1960, Glennan again made a plea to the Army to let the space agency have the von Braun team.

At the time, the team was developing the Pershing battlefield missile, and NASA and the Army worked out an agreement in which part of the group would stay with the Army until the Pershing flew successfully. Glennan also promised to help the Army form a new missile group. President Eisenhower's transfer order became effective May 15, 1960, and the von Braun rocketmen didn't have to leave Huntsville. NASA established the new George C. Marshall Space Flight Center there, appointed von Braun its first director, and named former Germans to all key positions. The team by then had grown to more than 4,000 with the addition of American engineers, technicians and managers.

"The move to NASA was a great relief," von Braun said. "No longer would I or my colleagues be producing rockets to kill people. For the first time we could plan for our long-held dream of exploring space."

Von Braun's immediate job was to develop the Saturn 1 rocket. "If we tried to develop a single rocket with that much thrust, it would have taken many years, and time was important if the United States was to overtake the Soviets," he said. "The wisest course at the time was to tie together several proven rockets, while at the same time planning for building even larger rockets of the future."

But it was von Braun's smaller Redstone that made the early headlines. On January 31, 1961, one of these boosters hurled a Mercury capsule carrying a chimpanzee named Ham on a suborbital flight more than 100 miles high — to the edge of space and back — in a dress rehearsal for America's first manned flight. But a combination of malfunctions associated with the Mercury safety equipment gave Ham a rougher ride than expected and he landed 422 miles downrange, 150 miles beyond the splashdown zone. By the time helicopters from a recovery ship found the bobbing spacecraft, it was on its side, had taken on about 800 pounds of water and was dangerously close to sinking. But a very annoyed Ham had survived the 15-minute flight ordeal, and he was rewarded on the ship with an apple.

The malfunctions made certain NASA engineers nervous, and they delayed the first Mercury astronaut flight, which had been set for March, so they could work out the bugs with one more test launch of the rocket and capsule. The extra test was completely successful, but that cautionary move probably cost America a chance to launch the first man into space.

This 1960 U.S. Naval Observatory time-lapse exposure of the Milky Way Galaxy shows the diagonal trail of Echo I in the foreground, a great distance from the backdrop of countless stars. Echo I, the first communications satellite launched by NASA, was a 100-foot sphere also used for air density experiments. It was operational almost eight years.

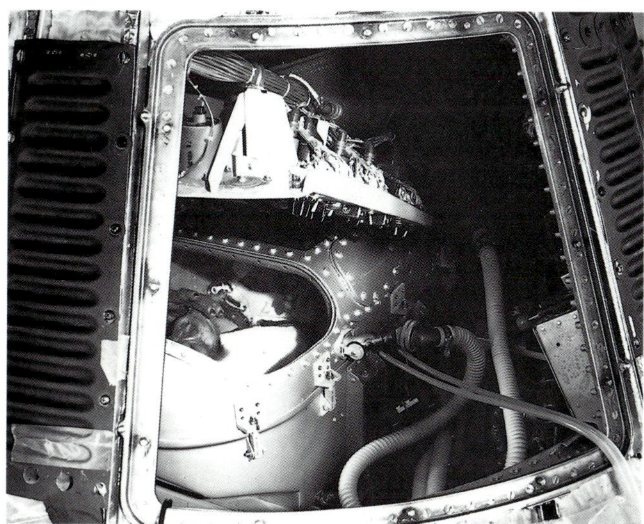

On January 31, 1961, a chimpanzee named Ham, shown left in his suit fitted into the couch of the Mercury capsule and right in the Mercury capsule just before launch, was hurled on a suborbital flight more than 100 miles high by von Braun's Redstone rocket.

Alan Shepard, the first American to fly in space, is shown here in his Mercury pressure suit.

Chapter Three

'Now Let the Other Countries Try to Catch Us'

On April 12, 1961, a warm, sunny day, a mighty booster rose from the Soviet launching base at Tyuratam amid the flower-covered steppes of central Asia. And, for the first time, a voice was heard from a rising rocket. "Poyekhali!" ("Let's go!"), and then, "The machine is working normally." The launch occurred at 9:07 a.m. Moscow time. At 10 a.m., the Soviet Union electrified the world by announcing that one of its cosmonauts, Air Force Maj. Yuri A. Gagarin, 27, was orbiting the Earth as far out as 203 miles in the new five-ton Vostok spacecraft which he had nicknamed *Swallow.* At noon, officials reported that he had returned safely to Earth, parachuting into a farmer's field, after circling the globe one time in 108 minutes.

The news hit America like a bombshell. The world's newest hero, the first man in space, was a Soviet citizen. It was hard for Americans, so used to success, to accept once again. Said Chris Kraft, Mercury flight control director: "That was a very bad, bleak day for us in the space program. We were unhappy that they had been able to beat us. They probably started at about the same time we did, and they probably didn't take any more risks than we did, and yet they beat us, and they beat us soundly."

Astronaut Alan Shepard, who had been selected to become the first American in space, was more than miffed. He had been angered at the decision to fly an extra unmanned Redstone-Mercury flight that probably forfeited his chance to best the Russians, and he made that clear when he reacted: "We had 'em, and we gave it away. I could have been up there three weeks ago."

Shepard reluctantly joined his fellow astronauts and his president, as well as heads of state around the world, in praising the flight.

Details of Gagarin's trip came out later. He had been subjected to six times the force of gravity during liftoff and up to 10 G's when Vostok slammed back into Earth's atmosphere. Flight surgeons had been concerned that he might become disoriented in weightlessness, but Gagarin said it had not bothered him. "At first it was a strange feeling," he reported. "But I soon got used to it and could work normally. Everything was suddenly easier. I felt as if my arms and legs, in fact my entire body, no longer belonged to me. I am not sitting or lying down. I seemed to hang somewhere in the air. All unsecured objects were floating around and when I looked at them I seemed to be dreaming. There flew the atlas, the pencil, the notebook."

Feeling good a day after his flight, Gagarin boasted: "Now let the other countries try to catch us."

Premature engine cutoff at launch of the unmanned Mercury Redstone 1 on November 21, 1960, terminated the test. The emergency escape system was jettisoned, but the spacecraft was not damaged. Despite this early setback, Mercury-Redstone proved a successful program for America's suborbital flights, lofting a chimpanzee named Ham and the first two U.S. astronauts into space.

Indeed, Gagarin's flight had widened the space gap. That was especially bothersome to America's young new president, John F. Kennedy, who had taken office three months earlier with a pledge to "get this country moving again." He had made the so-called missile gap and the space gap major issues in his campaign against Eisenhower's vice president, Richard M. Nixon. During the campaign, he had declared that "space is our great new frontier." Yet, Kennedy did little to close the gaps in his first months, concentrating instead on other problems like a sluggish economy, communist incursions around the globe and Fidel Castro's takeover of Cuba.

Gagarin's flight changed his outlook. Two days later, Kennedy summoned to the Oval office Vice President Lyndon Johnson; James E. Webb, NASA's new chief, and Jerome Weisner of Massachusetts Institute of Technology, who was the new presidential

NASA Administrator James E. Webb.

science adviser. Kennedy asked: "If someone can tell me how to catch up, there's nothing more important."

Weisner suggested that the U.S. stop advertising Mercury as its major space objective. He told Kennedy the nation had taken the lead in science, communications and military satellites and suggested he concentrate on these programs and play down, even forfeit, what could be a losing race to put men in space.

But Kennedy, Johnson and Webb reasoned that men, not robots, would capture the imagination of a world in which people and nations were choosing up sides and technical achievements were equated with power. There had been some discussion about eventually sending men to the moon, and Kennedy asked: "Can we go to the moon before them? Can we put a man on the moon before them?" With America having no manned space flight experience, it was a bold question indeed.

American prestige was rocked again three days later. On April 17, a brigade of Cuban exiles, sponsored by the U.S. Central Intelligence Agency, stormed ashore at Cuba's Bay of Pigs in an attempt to overthrow Fidel Castro. Castro's forces routed the ill-equipped exiles, who claimed the CIA failed to provide promised air cover. The failed invasion left the U.S. and its young president in shame and disarray.

Kennedy accepted the blame, and decided bold action was needed to restore America's honor. Space was the answer, he concluded, and he gave Vice President Johnson the task of determining what could be done.

After a series of meetings with officials of NASA, the Defense Department, the Budget Bureau and private industry, Johnson concluded there was a strong catch-the-Russians mood in the country, and in a preliminary April 28 report told the president a manned lunar landing was both desirable and technologically feasible. And it was something the United States had a chance of doing first.

Even as the president and his advisers were discussing going to the moon, there were some in his administration who were expressing concern about the first U.S. man-in-space flight by Alan Shepard, now scheduled for May 2. With everything else going wrong for the nation, the impact of a failure would be devastating.

Kennedy held an urgent meeting in the Oval Office on April 29 to discuss what should be done. Some suggested that he order the Shepard flight delayed. But Kennedy overruled them, saying it was time to bite the bullet and send a man into space. He let the launch date stand.

Chris Kraft's launch team counted down Shepard's Redstone on May 2, but the astronaut never made it to the pad that day. Heavy clouds moved over Cape Canaveral, and officials decided to wait for a better day.

Skies were clear on May 5, and at 5 a.m., the 37-year-old Navy lieutenant commander, clad in a silvery space suit, arrived at the launch pad. "I was struck by the beauty of the Redstone topped by the Mercury capsule," he said later. "The rocket, bathed in searchlights, was puffing white clouds from its liquid oxygen vents, and frost covered its sides. I told myself I was never going to see this rocket again, so I stopped and looked at it for a few seconds." He rode an elevator up the side of the Redstone and slipped into the spacecraft he had named *Freedom 7* — the seven representing the seventh Mercury capsule off the assembly line as well as the seven Mercury astronauts.

May 5, 1961: Mercury-Redstone 3, with Alan Shepard aboard, was launched on a suborbital mission from Cape Canaveral.

Shepard had to wait out more than four hours of technical and weather delays before the countdown finally reached zero at 9:34 a.m. and the Redstone burst to life, rising slowly at first as it hungrily consumed its tons of fuel. Shepard's vision became blurred momentarily as his body was pressed back against the couch by acceleration forces seven times the grip of gravity. The rocket burned out, dropped away and *Freedom 7* and its passenger were crossing over the threshold of space. For the next five minutes the astronaut was in a weightless world, enjoying a sensation that only one other, Yuri Gagarin, had experienced. "What a beautiful sight!," he exclaimed as he viewed the Atlantic and much of Florida through a small porthole.

America seemed to be rising with him. Millions crowded around radios and television sets to follow the progress, hanging on Shepard's every word, even though most of it to them was meaningless pilot talk. President Kennedy and his National Security Council interrupted a meeting to watch on television.

Freedom 7 accelerated to 5,180 miles an hour, but quickly slowed as the capsule plunged back into the atmosphere and gravity forces once again worked on the astronaut's body. The spacecraft soared 116 miles up and 303 miles downrange, parachuting into the Atlantic near Grand Bahama Island. A Marine helicopter retrieved Shepard from the floating capsule and carried him to the deck of the aircraft carrier U.S.S.

A U.S. Marine helicopter plucked Shepard from the Atlantic.

Champlain. "Boy, what a ride!" he told the crew. To the doctors, he said, "I don't think there's much you'll have to do to me." His only complaint, he said, "is that the flight wasn't long enough."

America heaved a great sigh of relief. The Russians were still far ahead, but the United States was on the move. Congratulations poured in, and there were many plaudits for the open way the U.S. had conducted the flight in contrast with the Russian secretiveness. Soviet Premier Khrushchev sent a salute. But earlier that day in a speech in Armenia, he downplayed Shepard's 15-minute flight compared with Gagarin's orbit.

Shepard became a national hero on the level of Charles Lindbergh, feted at the White House and honored with huge parades in Washington and New York.

President Kennedy awarded NASA's Distinguished Service Medal to Shepard in a Rose Garden ceremony May 8, 1961.

The tremendous impact of Shepard's flight on a nation thirsting for success was not lost on Kennedy, and he said: "All America rejoices in this successful flight.... But America still needs to work with the utmost speed and vigor in the further development of our space program." That's precisely what the president was doing, although the country did not know about it. Three days after Shepard's trip, Vice President Johnson presented Kennedy with his final report. NASA, Defense Department and other experts had concluded that landing a man on the moon was technically feasible, but the cost would be high — at least $20 billion. The president winced at the price tag. He was told by his science adviser, Jerome Weisner, that for one-tenth the cost and at no risk to life, instruments could be sent to the moon and learn as much as men could learn. But again Kennedy argued that men, not robots, would capture the attention of the world.

On May 25, 1961, four months after taking office, Kennedy stood before Congress and delivered a special State of the Union message on "urgent national needs." It covered many subjects, from the economy to disarmament, but its main thrust was space.

"Now is the time to take longer strides — time for a great new American enterprise — time for this nation to take a clearly leading role in space achievement, which in many ways may hold the key to our future on Earth.... Recognizing the head start obtained by the Soviets with their large rocket engines...and recognizing the likelihood that they will exploit this lead for some time to come in still more impressive successes, we nevertheless are required to make new efforts of our own. For while we cannot guarantee that we shall one day be first, we can guarantee that any failure to make this effort will make us last....

"If we are to win the battle that is going on around the world between freedom and tyranny, if we are to win the battle for men's minds, the dramatic achievements in space which occurred in recent weeks should have made clear to all of us, as did Sputnik in 1957, the impact of this adventure on the minds of men everywhere who are attempting to make a determination on which road they should take.... We go into space because whatever mankind must undertake, free men must fully share...."

Three weeks after Shepard's successful suborbital flight, on May 25, 1961, President Kennedy, addressing a joint session of Congress, committed the United States to landing a man on the Moon and returning him safely to Earth before the decade (1960s) ended.

Then the shocker.

"I believe this nation should commit itself," the president said, "to achieving the goal, before this decade is out, of landing a man on the moon and returning him safely to Earth. No single space project in this period will be more impressive to mankind, or more important for the long-range exploration of space; and none will be so difficult or expensive to accomplish.... I believe we should go to the moon, but I think every citizen of this country, as well as the members of Congress, should consider the matter carefully in making their judgment."

For Kennedy, his bold decision was a strategic and political one aimed at making America second to none, to advance its interests around the globe and to create a sense of accomplishment and pride at home.

Many in NASA thought Kennedy had lost his mind. How could he make such a commitment with only Shepard's 15-minute popgun flight as a guidepost, they asked. Said astronaut Gordon Cooper: "It just isn't possible. First, we don't have the rockets. Second, we don't have the spacecraft, and third, we don't even know how to navigate our way out there and back."

The project was called Apollo — after the Greek archer-god, the god of light and truth, bright-shining, far-shooting. America's largest, costliest and most ambitious technological effort began with virtually no dissent from the people or the Congress. Before it was done, more than 20,000 industrial contractors and 400,000 technicians, engineers, managers and other skilled persons would be involved in the massive enterprise.

The effort was well-managed. One of NASA's first decisions was to reject the idea of creating manufacturing facilities owned and operated by the government. Instead it farmed out most of the work to companies like aerospace giants McDonnell Aircraft, Douglas, North American Aviation, Martin, Boeing, Aerojet General, Chrysler and Lockheed, and thousands of smaller firms. By awarding contracts in nearly every state, the agency gained wide political support.

The task of developing the mammoth rockets for the project went to von Braun and his rocketeers at Huntsville. "We were overjoyed with the president's decision," von Braun said. "We were finally going to realize our hopes for space exploration."

But in 1961, the moon was a long way off. Many steps had to be taken, many hurdles overcome. And there was much more to learn from the astronauts of Project Mercury.

Chapter Four

'Man is the Deciding Element'

The second Mercury manned flight was to be a duplicate of Shepard's 15-minute suborbital jaunt to further verify the spacecraft design. The pilot was Virgil I. Grissom, a quiet, short man everyone called "Gus."

There were engineering changes in the capsule, including an improved hand controller and a window instead of a small porthole to give the astronaut a better view of the Earth. And, in a change that was to have a significant bearing on the flight, engineers wired the hatch so that it could be blown with an explosive charge, a quicker way of opening it than removing a series of bolts.

Grissom blasted off on July 21, 1961, in the spacecraft he had named *Liberty Bell 7*. Everything went well, and Grissom parachuted safely into the Atlantic Ocean. While bobbing in the water, reviewing his recovery checklist while waiting for a helicopter to pick him up, Gus was suddenly startled by an explosion that blew out one side of the capsule. The explosive primer cord had inexplicably ignited and blown away the emergency exit hatch.

Water began pouring in and Grissom had to scramble through the hatch and swim for his life as he watched the 3,000-pound *Liberty Bell 7* sink in 15,000 feet of ocean.

Engineers were mystified by the explosion. Some theorized the astronaut may have inadvertently hit an emergency plunger that triggered the hatch release. Gus denied this, repeatedly insisting, "The damn thing just blew." His fellow astronauts backed him all the way, and an accident review board cleared him of any wrongdoing. With the evidence on the ocean floor, it was impossible to determine what went wrong.

If the loss of *Liberty Bell 7* wiped some of the shine off an otherwise perfect flight, the rest of the luster vanished 16 days after Grissom's splashdown when another powerful booster raced skyward from its pad in Russia with Vostok II — *Eagle* — carrying Air Force Maj. Gherman S. Titov. He went aloft in a ship weighing nearly 11,000 pounds and remained in orbit a full day before returning safely home. His only problem in flight was a recurrent dizziness, especially severe when he turned his head sharply or watched fast-moving objects zip by the spacecraft window. To U.S. flight surgeons, the report raised a question of whether long-term weightlessness posed a danger to humans.

NASA managers and astronauts could only look on with envy. They agreed the Redstone had done its job. No more suborbital flights. It was time to get on with launching an American into orbit, and suddenly the nation's most urgent goal was to send a man circling the globe before the end of 1961. That's the year the Soviets did it, not once, but twice, and they did it first. But history could record that both nations had done it in the same year.

Astronaut Gus Grissom, shown here during egress training in June 1961, was America's second man in space.

While NASA put its full energy behind that goal, many observers said it could not be done. How, they asked, was the agency going to resolve the problem with the Atlas, the only rocket in the American arsenal capable of lifting the 3,000-pound Mercury capsule into orbit? The Atlas worked fine as an intercontinental range war rocket but it fizzled often under the burden of the Mercury capsule and all the redundancy and other safety features added to protect human life.

Three times the Atlas had lifted off with an unmanned Mercury craft on top, and twice the vehicle had exploded. Many systems were modified, the fragile metal skin was strengthened to support the capsule, and on September 12, five months after the last blowup, NASA was ready to try again. This time the rocket worked perfectly. An unmanned capsule was hurled into orbit and returned to Earth. In November, a chimpanzee named Enos rode an Atlas into orbit to blaze the trail for an astronaut to take the next trip.

Selected to make that journey was John Glenn, a freckled, boyish-faced Marine who had downed three Communist MIGs in the Korean War, who won five Distinguished Service Crosses and 18 Air Medals and

Enos, the chimpanzee that orbited the Earth twice in a Mercury spacecraft, arrives back at Patrick Air Force Base on November 30, 1961. Enos landed some 220 nautical miles south of Bermuda and was picked up by the U.S.S. *Stormes.* His medical debriefing at Bermuda showed he had no ill effects from his space ride.

who set a cross-country supersonic speed record in 1957.

NASA set a launch date of December 20. The date was premature. There were problems here, uncertainties there, and the flight slipped into January. History would record that only Russians orbited the globe in 1961.

Nine times NASA set a launch date for John Glenn, and each time it had to be delayed — mainly by stormy winter seas in the Atlantic landing and recovery areas. The country began to agonize for Glenn. Some suggested the psychological pressures of all the delays might be too much and that another pilot should be substituted. One of President Kennedy's advisers suggested the launch be put off until spring, when better weather could be expected. But Kennedy replied: "I know it is a strain on Colonel Glenn.... But I have taken the judgment that the judgment of those on the spot should be final.... I think they would be reluctant to have it canceled for three or four months because it would slow our whole space program down at a time when we are making a concentrated effort in space."

The storms finally subsided, and February 20, 1962, was the tenth date set for the liftoff. At 6 a.m., Glenn, wearing his silver space suit, once again rode the elevator 12 stories up the side of the Atlas and climbed aboard his *Friendship 7* spacecraft.

Nagging technical problems stalled the countdown, but they were resolved, and at 9:47 a.m., the Atlas roared to life. Tens of thousands who had gathered in the Cape Canaveral area, lining beaches, highways and riverfronts, watched fascinated as the 125-ton rocket thundered upward, spewing a trail of fire as it thrust man and machine into the sky. "Go, baby, go!" they shouted. "God speed, John Glennn," said a voice from Mission Control as the fiery rocket climbed higher and faster.

Across America, there was a hush, as people gathered before television and radio sets to follow the historic liftoff. President Kennedy and congressional leaders interrupted a breakfast meeting at the White House to watch TV.

"It's a little bumpy along here," Glenn radioed as the Atlas vibrated on liftoff. But the two booster engines and the main engine worked perfectly, and minutes later *Friendship 7* was in orbit 125 miles above the Earth, flashing through the sky at 17,400 mph.

"Oh, that view is tremendous," the astronaut exalted. "It looks beautiful...I can see clear back, a big cloud pattern, away back toward the Cape." He marveled at sights. The spectacular sunrises and sunsets, one every 45 minutes. The colors. The Earth below, the clouds in great white swirls. Camera, pencil and notebook floating before his eyes.

John Glenn and America were in orbit. Cheers erupted across the land.

The flight was planned for three orbits, each lasting about 90 minutes. Near the end of the first circuit, Glenn reported that a problem with a small steering jet was causing the nose of the spacecraft to drift slightly to the right while he was on automatic pilot. "I am on fly-by-wire now. I am controlling manually," he said. Mission Control recommended he continue manual control to conserve fuel. Officials said later that if there had not been a man on board to override the faulty automatic system, the capsule would have been commanded home after the first orbit.

Astronaut John H. Glenn, Jr., shown here at Cape Canaveral in his Mercury pressure suit, made America's first manned Earth-orbital space flight on February 20, 1962. The flight lasted 4 hours 56 minutes and covered 81,000 miles.

"God speed, John Glenn," said a voice in Mission Control as Mercury-Atlas 6 lifted off, carrying Glenn into orbit 125 miles above Earth.

The space flier adjusted well to his weightless environment, called it a pleasant experience. He reported none of the nausea or disorientation that had disturbed Soviet cosmonaut Gherman Titov the previous summer.

Four hours into the flight, on his third pass over the Pacific, Glenn began counting down for the firing of the braking rockets that would slow his speed and drop *Friendship 7* out of orbit. But for nearly an hour, Mission Control had worried about a warning light on a ground control panel. It indicated that the ship's heat shield might be loose. Without the shield, Glenn and his capsule would be burned up by the 4,000-degree heat that would punish the craft during a blistering plunge through Earth's thickening atmosphere.

What could be done if the heat shield were loose? Experts on the ground decided not to jettison the housing for the retro-rocket package. This was a pack of three small rockets strapped to the blunt end of the capsule, below the heat shield. Normally the straps would be broken by radio command from the astronaut and the pack would drop away after the rockets were fired to slow the vehicle for reentry. Operations director Walt Williams and flight director Chris Kraft reasoned that if the pack were left on, the straps would hold the heat shield in place at least until increasing air pressure at lower altitudes could do the job.

Glenn was told to keep his retropack on. But he was not informed why until he was almost back on Earth. He complained later that he should have been told earlier. But Chris Kraft said officials felt fairly confident the heat shield really was not loose, that it was a false signal, and they did not want to unduly alarm the pilot.

"I feel like I'm going back to Hawaii," Glenn ex-

A camera aboard *Friendship 7* captured this picture of John Glenn during the Mercury-Atlas 6 space flight.

claimed as the retropack fired with a triple jolt over the eastern Pacific. *Friendship 7* plunged into the atmosphere, and Glenn saw burning chunks of the rocket pack flash past his window in the fiery-orange glow of the heat shield, burning away as it absorbed heat and dissipated it into space.

He worried for awhile that the flaming objects might be chunks of the shield, but he felt no heat buildup and breathed easier. An envelope of ionized gases surrounded the capsule, momentarily blocking communications. When Glenn emerged from this blackout, he reported, with a great tone of relief, "Boy, that was a real fireball."

Friendship 7 parachuted into the Atlantic near Bermuda, within view of the recovery destroyer U.S.S. *Noa*. Twenty minutes later, the ship's crew hoisted the spacecraft, with Glenn still inside, onto the deck. The astronaut blew the hatch, stepped onto the deck, acknowledged the cheers of scores of sailors, and reported to doctors, "My condition is excellent."

The flight had lasted 4 hours 56 minutes and covered 81,000 miles.

America's first orbiter became a national hero as his countrymen reached out, in relief, pride and gratitude. Congratulations poured in from around the world. Glenn was honored at the White House, he was cheered by a quarter million persons in the rain in Washington, by four million in a ticker tape parade in New York, and by 50,000 who jammed into New Concord, population 2,000, the small Ohio town where he grew up.

The flights of Shepard, Grissom and Glenn had put muscle into President Kennedy's space race challenge to go to the moon. But Mercury was still cutting its teeth, and the moon was a long way off.

NASA moved quickly to a second Mercury orbital flight, with "Deke" Slayton set to be at the controls. But a month before he was to lift off, Slayton was grounded by flight surgeons who detected an irregularity in his heartbeat. He was replaced by Scott Carpenter. The other Mercury pilots talked officials into naming Slayton chief of the astronaut office, a desk job, hoping he somehow could stay around and regain flight status.

Carpenter's *Aurora 7* spacecraft rocketed into space on May 24, 1962, on a planned three-orbit trip, and the astronaut expanded on Glenn's role by conducting some science experiments and photographing sun, stars and Earth, in addition to testing the Mercury systems.

Carpenter inadvertently left two control systems on at one time, using up valuable fuel. And nearing the end of the mission, he fell behind in his checklists. When he fired his retrorockets he was a second late and was not holding the spacecraft at a precise angle. As a result *Aurora 7* landed in the Atlantic 250 miles beyond the target zone, out of sight and out of radio range. For 33 minutes he was virtually lost to Mission Control. A recovery aircraft picked up his homing radio beacon and found Carpenter floating in a life raft attached to the bobbing capsule, eating candy and snacks from a survival kit.

Mercury orbital flights originally were to have been limited to three orbits, about five hours. But Carpenter's mission, despite its problems, provided information on how to streamline the capsule by removing unnecessary heavy equipment and replacing it with more fuel and other consumables. NASA announced the next flight, with Wally Schirra in command, would be six orbits.

But before Schirra could fly, the Soviets, their manned space program idle for a year, demonstrated they were still well ahead. In August 1962, two cosmo-

Scott Carpenter, shown here in a procedures trainer in the Mercury Control Center at Cape Canaveral, made three orbits of the Earth on May 24, 1962, in *Aurora 7*.

Walter M. Schirra, Jr., pilot of the *Sigma 7*, made six orbits of the Earth on October 3, 1962.

nauts in separate spacecraft, Andrian Nikolayev in Vostok 3 and Pavel Popovich in Vostok 4, were launched a day apart. Initially the craft were just three miles apart, but gradually drifted to a separation distance of about 300 miles. They maintained radio contact throughout and conducted similar experiments for comparison purposes. Popovich orbited 48 times in 70 hours, Nikolayev 64 times in 94 hours. Schirra's upcoming nine-hour journey suddenly seemed very small.

Small, but good. An Atlas propelled Schirra's *Sigma 7* capsule into orbit on October 3, 1962. He conserved fuel by allowing the craft to drift through space for long periods, passing the time by snapping hundreds of photos. The technique worked well, and at flight's end, he had 78 percent of his fuel left in both the automatic and manual control systems. There were no problems and NASA called it a "textbook" mission.

Among those who watched the flight with intense interest were nine test pilots, the second group of astronauts enlisted by NASA. Some of them would become legends, and two would die for the cause. They were Neil Armstrong, Frank Borman, Charles Conrad, James Lovell, James McDivitt, Elliott See, Thomas Stafford, Edward White and John Young.

Schirra had demonstrated that the Mercury craft could stay aloft much longer than anyone could have imagined years earlier. So officials set the next flight for 22 orbits — 34 hours. Gordon Cooper got the assignment and his *Faith 7* lofted into space on May 15, 1963. "Working just as advertised," he reported as he settled into orbit. He snapped photos, collected urine samples for the medics, relayed the first television pictures from a manned U.S. spacecraft and attached sensors that recorded sleep patterns of the first American to doze in space.

All went well until the 19th orbit when a serious electrical problem knocked out *Faith 7's* position instruments and automatic control system. That meant Cooper had to manually fire his retrorockets and guide the capsule through the searing heat of reentry to a landing. He handled the tense situation perfectly and popped the spacecraft into the ocean just four miles from the main recovery carrier.

Cooper's flight was the last planned in the Mercury program. But NASA pondered whether to prolong a program that already had exceeded its original goals. Noting that the first flight of the two-man Gemini spacecraft was two years away, some suggested keeping Mercury alive with a 72-hour mission. But the target was the moon, and the decision was made to concentrate on the next step, Gemini.

Gordon Cooper, pilot of *Faith 7*, made 22 Earth orbits in 34 hours, and manually guided the spacecraft through reentry after an electrical problem knocked out position instruments and the automatic control system.

President John F. Kennedy accepts a model of an Apollo spacecraft command module from MSC Director Dr. Robert R. Gilruth, right, during a September 12, 1962 visit to the Manned Spacecraft Center. Looking on is Vice President Lyndon B. Johnson. Speaking to a crowd of 35,000 at Rice University Stadium in Houston that day, Kennedy said, "We intend to become the world's leading spacefaring nation."

Chris Kraft (front right), shown in this 1965 photo with NASA Administrator James Webb (front left), answered when asked the most important lesson learned from Project Mercury: "Man is the deciding element. As long as man is able to alter the decision of the machine, we will have a spacecraft that can perform under any known conditions, and that can probe into the unknown for new knowledge."

November 11, 1966: Mercury operations director Walt Williams, left, and astronaut Walter Schirra reminisce in front of the Mercury Monument at Pad 14, Cape Kennedy, Florida. The Arabic number 7 represents the seven original astronauts. The other figure is the astronomical symbol of the planet Mercury. In the background is the Gemini XII Agena Target Docking Vehicle atop its Atlas launch vehicle.

In June 1963, NASA closed the book on Project Mercury. Asked the most important lesson learned, Chris Kraft said, "Man is the deciding element. As long as man is able to alter the decision of the machine, we will have a spacecraft that can perform under any known conditions, and that can probe into the unknown for new knowledge."

A week after Mercury officially ended, the Soviets also launched the last of their one-man Vostok capsules. Valeri Bykovsky rode Vostok 5 into orbit on June 14. Two days later, Vostok 6 soared into space and, for the first time, the world heard a female voice from space — that of Valentina Tereshkova, 26, who had been a spindler in a textile factory and an accomplished parachutist. "Hello, Earth, here is *Seagull,*" she radioed. Bykovsky welcomed her to orbit, "Congratulations on your jump into space."

The two moved to within three miles of each other but were in different orbital planes and slowly drifted several hundred miles apart. Both returned to Earth on June 19. Tereshkova's flight was a propaganda triumph that Premier Nikita Khrushchev had sought, and he boasted, "Bourgeois society always emphasizes that women are the weaker sex. That is not so. Our Russian woman showed the American astronauts a thing or two. Her mission was longer than all of the Americans put together."

Though Project Mercury had ended and Gemini and Apollo were yet to come, NASA's unmanned space program was on the move. At right is an artist's concept of the meteorological satellites Nimbus and Tiros launched in 1964. Tiros (center left), the world's first weather satellite, made pictures of the Earth's cloud cover from its position 400 miles in space, locating hurricanes and typhoons and aiding in weather analysis. Between August 28 and September 23, 1964, Nimbus I took 27,000 remarkably sharp daytime and nighttime pictures, including this one (center right) of Spain and the Mediterranean Sea, from its position 450 miles above Earth. Photo at bottom left was taken at Cape Kennedy Complex 37 August 25, 1964, during a Saturn 7 Radio Frequency Interference test for the Apollo program. The Pioneer A spacecraft, bottom right, carried six scientific experiments to study the interplanetary magnetic field, the solar wind and cosmic rays as it journeyed to within about 77 million miles of the Sun after six months of flight.

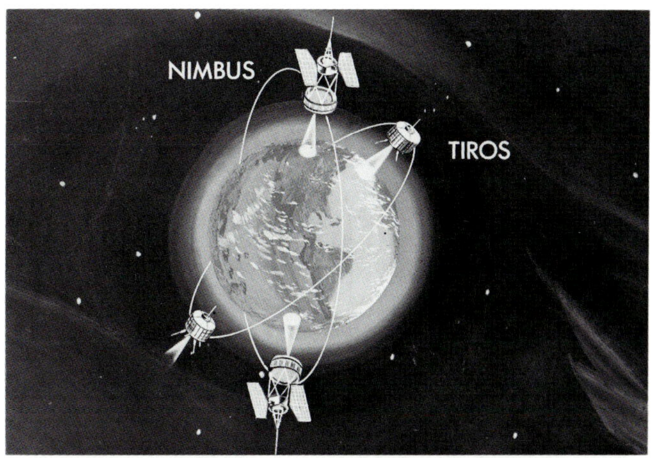

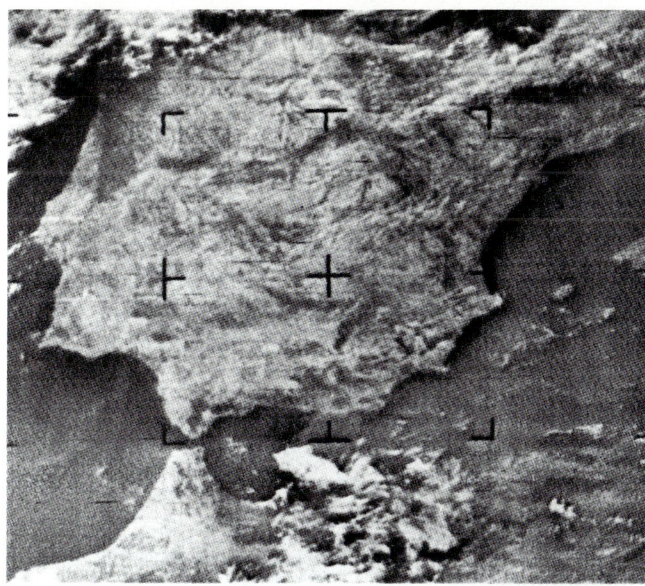

Astronaut Edward H. White II, pilot for the Gemini-Titan 4 space flight, floats in the zero gravity of space outside the Gemini 4 spacecraft. He is wearing a specially designed space suit, and the visor of his helmet is gold plated to protect him from the unfiltered rays of the sun. He also wears an emergency oxygen chest pack. He is secured to the spacecraft by a 25-foot umbilical line and a 23-foot tether line, both wrapped in gold tape to form one cord. In his right hand he carries a Hand-Held Self-Maneuvering Unit (HHSMU) with which he controls his movements in space. White egressed the spacecraft on the third revolution of the flight and spent 21 minutes outside. Astronaut James A. McDivitt, command pilot for the mission, remained inside the spacecraft. The first American to make a space walk, White's trip outside Gemini 4 came less than three months after Cosmonaut Alexei Leonov made the historic first space walk for the Soviets.

Chapter Five

'A Man is Floating Free in Space'

The husky cosmonaut, bundled in a bulky space suit, sat in a cramped airlock and cranked a handle that opened a hatch to the outside. "The hatch is open; I am getting out," he announced. His flying companion in the Voskhod 2 spacecraft radioed to the ground, "At this moment a man is floating free in space."

Alexei Leonov, Earth's first human satellite, reported he "popped out of the hatch like a cork" into a "flow of blindingly bright sunlight like an arc of electric welding." A television camera mounted on the side sent live pictures to the Soviet Union, dramatically depicting the historic space walk, on March 18, 1965.

For 10 minutes in time and about 3,000 miles in distance, the Air Force lieutenant colonel floated with his arms outstretched, tethered to the spaceship by a 15-foot lifeline and protected by his suit against space radiation and the extreme temperatures outside. He had no sensation of speed as he and Voskhod 2 raced along at more than 17,400 mph nearly 120 miles above the Earth, "I didn't experience fear," he said later. "Only a sense of infinite expanse and depth of the universe."

The time flew by. Leonov lingered for one last look at the Earth before he slid back into the airlock, feet first. He found the job more difficult than the exit and he had to partially deflate his spacesuit to squeeze back in.

Trouble loomed for Leonov and his commander, Pavel Belyayev, when it came time to fire the retro-rockets for return to Earth. The automatic system failed and Belyayev had to manually align the ship and fire the rockets. Off course and behind schedule, Voskhod 2 landed in deep snow in a forest in the Ural Mountains, 2,000 miles beyond the intended target. The spacemen spent a cold, uncomfortable night in the woods before they were rescued by a ski patrol.

In Moscow, a Soviet space official said cosmonauts will have to master space walking if they are to construct future space stations in orbit. He also spoke of the immediate goal: "The target before us is now the moon."

American space officials were surprised at Leonov's space stroll. They had no idea the Soviets were planning such a major milestone that early. "Coming just five days before our first Gemini flight, it caught us completely off guard," said NASA flight director Chris Kraft. The Americans felt the Soviets were concentrating at the time on perfecting the new Voskhod craft, which had flown for the first time five months earlier with the first three-cosmonaut crew, also a surprise.

It would be nearly four years before America's three-man Apollo ship would be ready to fly, and the Communist Party paper *Pravda* took note of this with a mocking headline: "SORRY APOLLO!"

Gemini-Titan 3, the first manned Gemini, lifts off March 23, 1965. The *Gemini 3* spacecraft, *Molly Brown*, carried astronauts Virgil I. Grissom, command pilot, and John W. Young, pilot, on three successful orbits of the Earth.

The Soviet space walk took some of the luster off the flight of *Gemini 3*, the first piloted flight of America's two-man spaceship and an important step in the effort to overtake the Russians.

Gemini was the bridge between man's first tentative forays into space with Mercury and the Apollo flights. The project was designed to investigate in actual flight all of the critical maneuvers that would confront the moon travelers — rendezvous and docking with another spacecraft, space-walking and long duration flight up to two weeks.

As in any new high-technology endeavor, there were problems developing the Gemini spacecraft as engineers and technicians strained the limits of the unknown. But by early 1964, most of the problems were resolved and the first Gemini, an unmanned test mission, lifted off on April 8 to determine performance and structural integrity of the spacecraft–Titan II rocket combination. After three orbits, the objectives were achieved. No recovery was planned and the vehicle burned up in the atmosphere.

Gemini 2, launched January 19, 1965, was an unmanned suborbital test. The flight proved the heat shield could protect the spacecraft under maximum reentry heating, and everything else went so well that it cleared the way for the first manned Gemini test two months later.

The two unmanned Geminis departed not from Cape Canaveral, but from its new name, Cape Kennedy — in honor of a president slain by an assassin's bullet who tragically did not live to see his moon dream fulfilled. Historians later convinced the Kennedy family to restore the original name, noting Cape Canàveral, named by Spanish explorers, was the first location ever to appear on a map of the New World. The NASA area of the Cape, where Apollo moon rockets would be launched, retained the name Kennedy Space Center.

Alan Shepard, America's first man in space, was to have commanded *Gemini 3*, but an inner ear problem that caused dizziness and nausea grounded him and he joined Deke Slayton in the astronaut office, both hoping for future cracks at space flight.

Gus Grissom, veteran of the second Mercury flight, and John Young, the first of the second group of astronauts selected for a mission, got the assignment. On March 23, 1965 — five days after Leonov's space walk — they raced into orbit atop a modified Air Force Titan II rocket.

Larger than Mercury, but small by Apollo standards, the 7,000-pound Gemini carried the two men in tight quarters about the size of the front seat of a compact automobile. The Titan II launch vehicle was the largest available at the time, and it could not manage a larger payload.

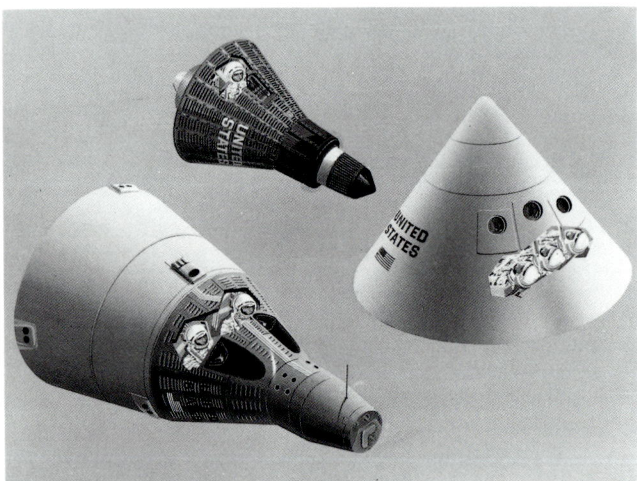
An artist's concept of the one-man Mercury spacecraft, the two-man Gemini spacecraft, and the three-man Apollo spacecraft.

Small or not, Gemini's systems worked well. On each of the three orbits, Grissom and Young fired the ship's steering jets, moving it from side to side, up and down, turning it around and changing its orbital course — the first manned spaceship, U.S. or Russian, to demonstrate this maneuverability. "It handles easily," Grissom reported.

The astronauts splashed down in the Atlantic 60 miles short of target because wind tunnel tests did not provide proper information on how much lift the craft would develop during reentry — a problem corrected for subsequent flights.

NASA had not planned a space walk until the *Gemini 6* flight late in 1965, but the schedule was advanced to *Gemini 4* after Leonov's surprise.

On June 3, 1965, two Air Force majors, James A. McDivitt and Edward H. White II, both from the second astronaut group, became the second Gemini team to orbit. The flight was planned for four days, with White's space walk set for the very first day.

McDivitt and White leave the suiting trailer at Pad 16 for the van ride to Pad 19 to board *Gemini 4* for their 4-day space flight during which White made his historic space walk.

White strapped an eight-pound container of emergency oxygen to his chest. He affixed a gold-tinted visor to the face plate of his space suit to reduce the sun's glare. The 21 layers of protective material in the suit would protect him from harsh temperature extremes of 250 degrees Fahrenheit on the sun side of the capsule and minus-215 degrees on the cold shadow side. He connected the gold-plated 25-foot tether that would be his link to the spacecraft and provide a communications channel.

High above Australia, four hours after launch, the astronauts began depressurizing their cabin, bleeding out oxygen until the inside and outside environments matched. Because of weight limitations, the Gemini craft could not afford the luxury of an airlock like that used by Leonov. Thus, both men of *Gemini 4* were exposed to the hostile environment of space.

Over the Pacific, between Hawaii and Mexico, White opened the hatch and slipped into the emptiness of space. He hurtled across the vivid panorama of Earth, more than 100 miles below, at 17,400 mph, but he had no sensation of speed. There were no trees or billboards in space to provide such a sensation.

White performed an out-of-this-world ballet, propelling himself around the spaceship with a handheld gun-like device that spat small jets of oxygen. He somersaulted, floated lazily, stood on Gemini's titanium hull.

The two astronauts were so intent and White was enjoying himself so much that the 12 minutes allotted to the space walk passed quickly. Capsule communica-

Astronaut Edward H. White II explains to President Lyndon B. Johnson how he used the jet gun with camera attached on his historic space walk during Johnson's visit to the Manned Spacecraft Center in Houston.

tor Gus Grissom, in Houston's brand new Mission Control, being used for the first time, ordered: "*Gemini 4*, get back in!"

White was reluctant. "This is fun. I don't want to come back in, but I'm coming," he said. But it was slow going and difficult for him to move without handholds or footholds in weightlessness, and it took White a few more minutes to get back into the cabin. He had been out 21 minutes. Then he had trouble closing the hatch, and McDivitt held him by the legs as he manhandled it shut. Said America's first spacewalker after he was back in: "It's the saddest moment of my life."

White later described his stroll. "There was absolutely no sensation of falling. There was very little sensation of speed...There were no fears...The view from up there is something spectacular."

McDivitt and White then settled down to the routine of flight. Their main goal during four days in orbit was to determine how humans would fare in weightless space for a prolonged period. When they returned to Earth, the medics examined them for three days and found them to be extremely healthy.

Officials were euphoric with White's space walk, but they were concerned about the difficulty he had in returning to the cabin. The general feeling was that whatever gave him problems could be easily resolved. Later space walks would demonstrate how wrong they were.

Buoyed by the medical data, NASA decided to double *Gemini 4's* time, and the August flight of *Gemini 5* was scheduled for an eight-day mission piloted by Gordon Cooper and Charles "Pete" Conrad. The spacecraft was plagued by problems with fuel supplies, power systems, a computer and venting gases that sent the vehicle into an annoying tumble. But the astronauts nursed their ship along and by the time *Gemini 5* came home, they had logged the longest manned space flight to date — 190 hours, 56 minutes, breaking the Soviet mark of 119 hours.

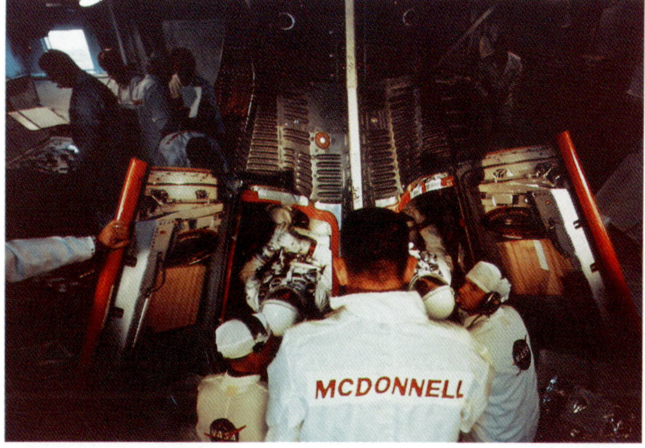

Gemini 5 astronauts L. Gordon Cooper Jr. (left), command pilot, and Charles Conrad Jr., pilot, practice making ready for liftoff inside their spacecraft with the aid of technicians from NASA and McDonnell Aircraft Corporation at Pad 19.

Chapter Six

'We're Tumbling End Over End'

In an out-of-this world sight, more than 100 miles above the Earth, two spaceships were doing fly-arounds, circling each other in lazy pirouettes, in sashay lefts and sashay rights. *Gemini 6* nudged to within one foot of its sister ship, backed off and came in again. *Gemini 7,* with a shortage of maneuvering fuel, made only a few twists and twirls. "There seems to be a lot of traffic up here," remarked commander Wally Schirra in *Gemini 6*. "Call a policeman," suggested commander Frank Borman in *Gemini 7*.

On Earth, it was December 15, 1965. In space, it had been a rapid sequence of sunrises and sunsets every 45 minutes as *Gemini 6* pursued *Gemini 7* in a 105,000-mile chase, caught up and flew in formation with it for five hours. It was the first rendezvous of two manned spacecraft in orbit, a test of a technique vital to U.S. efforts to land men on the moon and bring them back.

The rendezvous between *Gemini 6* and *Gemini 7* had not been in NASA's original plans. Schirra and Tom Stafford were originally to have chased down and docked their *Gemini 6* with an unmanned Agena satellite. But the Agena exploded on its way to orbit, and agency officials came up with the ingenious plan to launch *Gemini 7* astronauts Borman and Jim Lovell first on their record 14-day mission and send *Gemini 6* in pursuit.

It was another milestone reached on the road to the moon. But linking two ships in orbit still had to be proven. That job now fell to the *Gemini 8* crew of Neil Armstrong and Dave Scott. Their Agena target satellite, boosted by an Atlas rocket and its own engine, blasted into orbit on March 16, 1966, and the astronauts roared off in pursuit 90 minutes later on what would become Gemini's most harrowing mission.

Gemini 7 astronauts (right) are Frank Borman, top, command pilot, and James A. Lovell Jr., pilot. At left are backup crew members Edward H. White II, top, and Michael Collins.

Back on board rescue ship U.S.S. *Wasp, Gemini 6* pilots Walter Schirra and Tom Stafford cut a congratulatory cake.

Five hours after the chase began, the astronauts caught their prey, and Armstrong fired thrusters to brake his speed so it matched that of the Agena. Then he gingerly nudged *Gemini 8's* nose into a docking collar at one end of the Agena. Three clamps on the collar gripped the nose and an electric motor pulled it firmly into place. The clamps tightened, and for the first time two vehicles in space were physically and electronically linked as one rigid craft. "Flight, we are docked," Armstrong radioed. "It's really a smoothie. No oscillations at all."

The success was significant because it was a necessary part of returning two astronauts from the surface of the moon. They would have to guide their lunar module to just such a linkup with the Apollo mother

The Agena Target Docking Vehicle seen from *Gemini 8* during rendezvous in space.

Crewmen for the *Gemini 8* mission were astronauts Neil A. Armstrong, right, command pilot, and David R. Scott, pilot.

ship orbiting the moon. In Mission Control, there were cheers and hand shakes. But the jubilation was short-lived.

About 45 minutes after the linkup, over China and temporarily out of radio contact with Gemini ground stations, Scott noticed the joined spacecraft were beginning to move strangely, straying out of position. Almost instantly, the two vehicles "took off in roll and yaw," rolling like a log in water while the nose swung wildly.

The pilots felt something had gone wrong with the Agena controls, so they shut down the satellite's systems. Working skillfully with the Gemini controls, Armstrong, who had flown the X-15 rocket plane as a test pilot, managed to reduce the spin to a point where it was safe to pull free from the Agena. But after disengagement, the rotation got worse, spinning *Gemini 8* crazily at more than one revolution a second.

They were in the first real emergency in manned space flight.

Armstrong and Scott had thought the problem was with the Agena. But the fault lay with their own Gemini. One of the craft's 16 maneuvering thrusters had stuck open, spewing fuel into space and imparting the wild spin. They were close to the point where the ship would tear apart or they would fall unconscious.

A tracking ship, the Coastal Sentry *Quebec,* far out in the western Pacific, came into radio range and immediately detected the two craft had separated. An electrifying message came from Armstrong: "We consider the problem serious. We're tumbling end over end." His voice was calm, masking the anxiety of the situation.

The quick-thinking Armstrong decided they had one chance at survival — to shut down the primary thruster maneuvering system and switch to the nose rockets normally reserved to stabilize the ship during the dangerous and fiery reentry into Earth's atmosphere.

One by one, the 15 working thrusters of the primary system were shut down, and Armstrong turned on the reentry jets, using them to hold the vehicle as stable as possible until the bad thruster exhausted its fuel. "We are regaining control of the spacecraft slowly," he reported.

Control was regained after a hair-raising 32-minute struggle, but Armstrong and Scott would have to terminate their planned three-day flight as quickly as possible because continued use would deplete the fuel supply of the only system available for reentry.

Mission Control ordered Armstrong and Scott to make an emergency landing in the western Pacific. High over the African Congo, with no tracking stations to monitor their reentry path, they fired their retrorockets, and with great skill steered *Gemini 8* to a splashdown 480 miles southeast of Okinawa after a flight that lasted just under 11 hours. Thirty minutes later, an Air Force plane dropped pararescue men to secure the bobbing spacecraft, and in another three hours the astronauts were safely aboard a Navy destroyer.

An anxious Chris Kraft (left) watches the monitors in Mission Control with MSC Deputy Director James Elms.

Commented Chris Kraft: "Had it not been for that good flying, we probably would have lost that crew."

Rendezvous and docking had now been demonstrated but not yet perfected. That was done on the final three Gemini missions — 10, 11 and 12 — when the crews chased down and docked with Agena satellites, using the Agena engines to maneuver through

(Back) Astronauts Thomas Stafford and Eugene A. Cernan became *Gemini 9's* pilots after prime crew members Charles Bassett and Elliott See died in a plane crash.

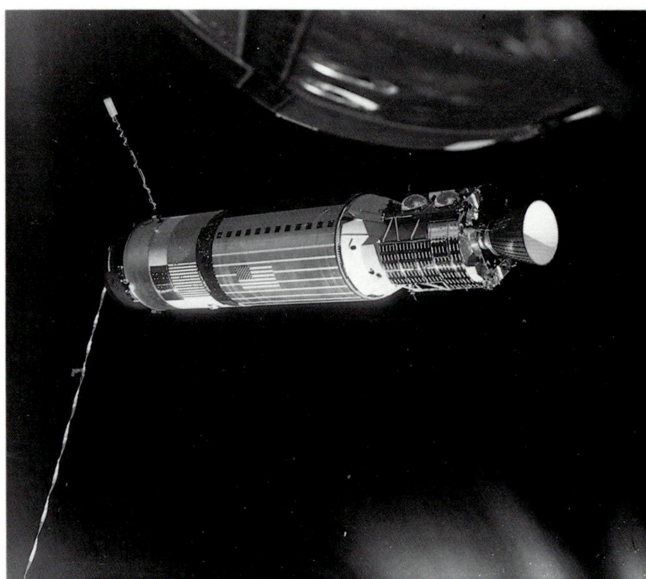

The Agena Target Docking Vehicle, photographed from the *Gemini 11* spacecraft, after separation. While docked, the two vehicles reached a new altitude of 740 nautical miles.

the skies. During the trackdowns, they rehearsed all the critical moves Apollo moonwalkers would have to make when they fired off the lunar surface to fly to a reunion with their mother ship.

Space walking became the most difficult Gemini nut to crack.

A year had passed since Ed White had taken America's first space walk. Officials should have been alerted by the trouble White had returning to his capsule, but because of his earlier ease in cavorting outside, they were fully confident that space walking was a piece of cake. They learned during *Gemini 9* just how wrong they were.

Gemini 9 started out as a jinxed mission when its prime crew, Charles Bassett and Elliott See, died in a plane crash. The backups, Tom Stafford and Gene Cernan, took their place and on June 3, 1966, they vaulted into orbit.

The next day, Cernan opened his hatch and climbed outside 185 miles up and pushed away from the craft on the end of a 25-foot lifeline. He moved to an open equipment section at the rear of the vehicle where he was to have donned a rocket-powered astronaut maneuvering unit (AMU) and hook up to an extra 125 feet of tether.

But the complexities of checking out and connecting with the AMU required much more work than expected. Cernan reported he had trouble just maintaining his position, continually having to use both hands to grip supports to prevent his floating away into space. He breathed heavily and perspired. Fog collected inside his helmet visor and froze.

Stafford was concerned. "How you doing, Gene?" he radioed. "Really fogged up, Tom," the spacewalker replied. Stafford told Mission Control, "It's about four or five times more difficult than we anticipated. And the pilot's visor is completely fogged over. Communications are very poor. He sounds like a loud gargle. If the situation doesn't improve..."

Stafford interrupted his own sentence: "It's no-go on the AMU! The pilot has fogged up completely."

"We confirm the no-go," Mission Control responded.

With great effort, Cernan returned to the cabin, working his way hand-over-hand along the side of the spacecraft, able to see through sections of the visor where his breath melted the frost. His heart beat at 180 ticks a minute — 100 above normal — as he worked feverishly to stow equipment before closing the hatch.

Future space walks with the AMU were canceled until working outside a spacecraft could be accomplished with a level of reliability and non-exhaustion. But problems persisted. Both Mike Collins on *Gemini 10* and Dick Gordon on *Gemini 11* experienced troubles maneuvering outside. Collins, who used a jet gun to move over to a nearby Agena satellite, reported: "I found that the lack of a handhold is a big impediment. I could hang onto the Agena, but I could not get

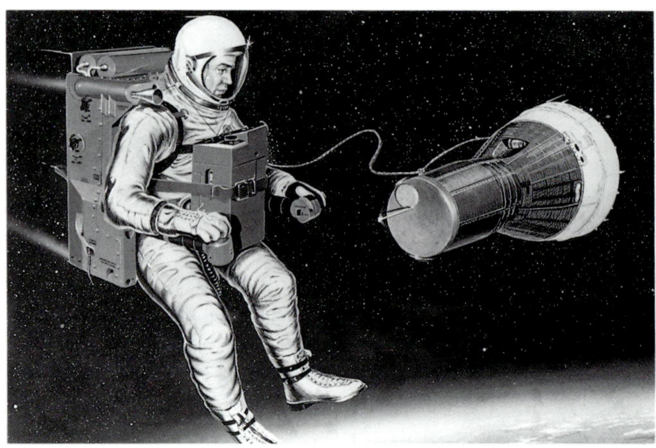

An artist's concept of how Cernan's spacewalk using the AMU was supposed to have worked.

Gemini 10 crew John Young, left, command pilot, and Michael Collins, pilot.

Gemini 12 astronauts James A. Lovell Jr., right, command pilot, and Edwin E. Aldrin Jr., pilot.

around to the other side where I wanted to go. That indeed is a problem." Gordon, like Cernan, became hot and sweaty and his visor fogged. "I'm pooped," he said simply after cutting his walk short.

Alarmed NASA officials asked: "How can we send men to the moon, if they can't work when they get there?"

One more Gemini flight remained in which to solve the perplexing problems of body positioning and fatigue on three straight space walks. Veteran Jim Lovell would command *Gemini 12* and his space-walking partner would be Buzz Aldrin, who built on the experience of the others. When the astronauts left Earth on November 11, 1966, Aldrin took with him special devices like a wrist tether and a waist tether like those worn by window washers on high ledges. He carried portable handholds he could slap on to either the Gemini or docked Agena to keep his body under control. Mounted in a work station at the rear of the spacecraft were a pair of "golden slippers," foot restraints like Dutch shoes, that held his feet firmly in place as he operated easily at a work bench, loosening and tightening bolts, making tube connections and snipping wires.

Aldrin worked effortlessly for more than two hours, completing all his tasks without a problem. He demonstrated that a human being could work outside in space if he had proper equipment, hours of training and frequent rest stops.

The astronauts returned to Earth on November 15, marking a triumphant end to Project Gemini. In literally the last at bat, Aldrin had solved a critical problem that could have barred the road to the moon.

Gemini's 10 flights in 20 months had given the United States the lead in the space race. No Soviet cosmonaut had yet mastered rendezvous or docking. Only one space walk had been conducted. The best the Russians had done to date had been basic plodding in orbit compared with the free-wheeling flights of the Gemini crews.

President Johnson summoned Lovell and Aldrin and top space officials to his Texas ranch for a project-ending ceremony.

"Project Gemini has prepared, and prepared well, for the more ambitious Apollo flights that are to come," the president said. He added these words of caution: "The months ahead will not be easy as we reach toward the moon.... But with Gemini as a forerunner, I am confident we will overcome the difficulties and achieve another success."

Gemini 11 command pilot Charles Conrad Jr., right, and pilot Richard Gordon Jr.

Dr. Robert R. Gilruth (left) smokes a cigar in Mission Control to celebrate successful splashdown of *Gemini 11*. With him are MSC deputy director James C. Elms, center, and Gemini program manager Charles W. Matthews.

Chapter Seven

'I've Got a Fire in the Cockpit!'

Gus Grissom was angry. Nothing seemed to be going right with preparations for the first Apollo manned mission, which he was to command. He told an interviewer he had misgivings, that the flight had "a pretty slim chance" of going the full 14 days in Earth orbit. One morning, after a particularly bad performance by the training simulator at the Kennedy Space Center, he hung a lemon on it.

The picture seemed so rosy in late 1966 when the Gemini program ended as a huge success. Two unmanned Apollo spacecraft had been tested in space, hurled there by the powerful Saturn 1B rocket developed by Wernher von Braun's team for practice flights before the much larger Saturn 5 carried astronauts to the moon. Indeed, the road to the moon appeared to be a smooth one. Some in NASA were predicting Americans would be on the lunar surface as early as 1968, well before President Kennedy's deadline.

But with Gemini in the history books, the spotlight turned on Apollo, and the warts began to show. Apollo 1, the first manned flight, was set for November, just a month after the final Gemini mission. Then things started to go wrong with the Apollo command module, little things and big things. All costing time and money. Engines blew up on test stands. Trouble developed with a power inverter, an oxygen pressure regulator, the communications, water supply and life support systems. The launch slipped to December, then January, then February.

The mighty Saturn rocket made it possible for men to fly to the moon. Here, the mobile service structure for the Apollo 4 (Spacecraft 017/Saturn 501) unmanned space mission arrives at Pad A, Launch Complex 39.

There also were problems with the lunar module, especially with overweight parts and with the ascent engine intended to lift two men off the moon. But the lunar machine would not fly for awhile, and NASA concentrated on fixing the command module.

In early January, NASA set February 21 as the newest launch date for Apollo 1. On January 6, the spacecraft and its Saturn 1B rocket were moved to Complex 34 at Cape Kennedy. To save time, the agency cut a corner. It eliminated an unmanned test on the pad in which the command module was to have been pressurized with 100 percent oxygen for the first time. Instead, the cabin would be pressurized for the first time during an important countdown rehearsal with the astronauts on board.

Against this backdrop, the three men who were to fly *Apollo 1* went to the pad in their spacesuits for the countdown test on that fateful day, January 27, 1967. With Grissom were Ed White, America's first spacewalker, and a rookie spaceman, Roger Chaffee.

They had been scheduled to enter the craft at 11 a.m., but minor technical problems delayed their ride up the gantry elevator until 1:19 p.m. They strapped themselves in their couches, Grissom on the left, White in the middle and Chaffee on the right. Rocket and spacecraft, standing 224 feet tall, were encased in the steel latticework of a 310-foot service tower that provided access platforms for workers.

The rehearsal that day was called a "plugs-out" test. Its purpose was to check out the compatibility of the Apollo spacecraft and the ground support equipment and to evaluate communication and instrumentation systems and the ejection of electrical cables when the simulated countdown reached zero.

The hatch was closed and sealed and the cabin was pumped full of pure oxygen to a pressure of 16.7 pounds per square inch, two pounds higher than the normal oxygen-nitrogen sea level pressure outside the craft. The inside pressure was higher to keep contaminated air from seeping into the cabin. In space, the pressure would be reduced to 5 psi.

Throughout the count, the astronauts complained of a sour odor and poor communications, and several

Apollo 1 astronauts (left to right) Edward H. White, II, senior pilot; Virgil I. Grissom, command pilot, and Roger B. Chaffee, pilot, lost their lives in the January 27, 1967, fire in the Apollo command module during testing at Cape Canaveral.

holds were called. At one point, Grissom grumbled to the Cape control center, "If I can't talk to you five miles away, how can we talk to you from the moon?" With 10 minutes to zero, Grissom was unable to shut off his microphone by pushing a button and another hold was called. Someone suggested calling off the test. But time was important, pressures were high to get the program moving, and officials decided to go ahead, bypassing the mike problem. The count was to resume at 6:31 p.m. It never did.

A minute before that, beneath Grissom's couch, a bruised or broken electrical wire contacted exposed metal, shooting a shower of sparks into nylon netting. The netting, spread throughout the cabin to prevent loose objects from floating into crevices during weightless flight, started to burn. The flames spread rapidly in the 100 percent oxygen atmosphere.

A terrifying shout from Ed White blistered the communications circuit: "Fire!" Then a cry from Gus Grissom: "I've got a fire in the cockpit!" Followed by a garbled transmission and a final plea: "Get us out!"

In the control center, horrified watchers stared in disbelief at a television monitor as the flames exploded, became a terrifying white glare. In the extreme heat, the internal pressure of the cabin rose from 16.7 psi to 29 psi and the inner hull ruptured. Fire, gases and hot air rushed fiercely toward the opening, sweeping across the astronaut couches, feeding on the netting and other flammable material.

Solder joints melted, allowing streams of oxygen and an anti-freeze-like cooling fluid from the life support system to escape and feed the inferno. A wall of flames rose between the astronauts and their hatch, the only means of escape. Even if they could have reached it, it would have taken at least 90 seconds to open, and they all were dead within seconds after the blaze erupted.

"Asphyxia due to inhalation of toxic gases due to fire," was listed as the official cause of death. Because of the protective, non-flammable space suits, there were only minor burns on the bodies.

Surprisingly, the plugs-out test was not considered

hazardous and thus there were no safety personnel on the launch pad. Nor was there any equipment to fight this type of fire. In order for a test to be considered hazardous, there had to be fuel in the rocket, according to NASA rules.

There were 27 men working at various levels on the rocket tower. They grabbed gas masks and rushed to the spacecraft level. But the masks were designed against fumes from fuel spillage and were ineffective against the smoke. All 27 were treated for smoke inhalation. The heat and smoke prevented the workers from opening the badly charred capsule for five minutes.

Four days later the three astronauts were buried with full national honors — Grissom and Chaffee side by side at Arlington National Cemetery and White at the U.S. Military Academy at West Point.

A blue-ribbon board of review and congressional committees investigated the accident. Sharp questions were asked of both NASA and the company that built the spacecraft, North American Aviation. Why didn't the Apollo ship have a quick-escape hatch? Why was there so much inflammable material in the spacecraft? Why does the U.S. continue to use pure oxygen for breathing in spacecraft? Why were no emergency procedures in place on the launch pad? Why were there unprotected wires? Was there haste in the program to land men on the moon?

On April 9, the board of review, after 10 weeks of investigation, issued a 3,300-page report that weighed 19 pounds. Because of extensive damage, the cause of the fire could not be pinpointed precisely. The most likely source was given as an electric arc in defective wiring. Investigators found that insulation on wires under a small trap door in the environmental control system might have broken or frayed by rubbing against the metal door. Nine other possible sources were listed. All involved electrical failings in the same area, under Grissom's couch.

The report also criticized NASA and North American Aviation for poor management, carelessness, negligence, sloppy work and the failure to adequately consider the safety of the astronauts.

"The board's investigation revealed many deficiencies in design and engineering, manufacturing and quality control...numerous examples in the wiring of poor installation, design and workmanship," as well as poor welding and soldering of joints that carried flammable material throughout the spacecraft, the report said.

NASA and North American accepted the blame and answered a new call to arms to bring a safe Apollo on the line. There were numerous shakeups in NASA personnel, and most of the leadership at North American's space division was overhauled.

In the months ahead, nearly a half billion dollars would go into an exhaustive design and rebuilding of Apollo, including a new hatch that a man could open in three seconds. The new spacecraft also included extensive use of fire-resistant materials, a redesigned electrical system, better protection for plumbing lines, and the use of a combination oxygen-nitrogen atmosphere system when the spaceship was on the pad.

What happened to the command module also could happen to the lunar module, so extensive changes also were made in that spacecraft.

NASA had estimated after the fire that it would be a year before a redesigned Apollo would fly in Earth orbit with men aboard. But this time there was no rush. The job was done with painstaking care and attention to detail, and that year stretched to 21 months.

The accident generated another problem for the space agency. It sparked a national debate over whether the country should be spending an estimated $24 billion to send men to the moon. It was a time of growing opposition to the war in Vietnam, of campus unrest, of rising taxes and of concern for civil rights and the environment. Public opinion polls found more and more Americans asking: Is this trip really necessary?

The critics said the program cost too much, that the money could be better directed to problems on Earth, and that the race to the moon was just a political stunt. Many prominent scientists argued that less expensive unmanned probes could learn more about the moon than astronauts.

Apollo's defenders said the program would produce untold scientific and technological benefits, enhance national security and would demonstrate the nation's ability to lead in an age of technology.

Some of Apollo's technological spinoffs were making their way into everyday life. Most important were the compact, fast-acting computers designed to keep track of nine million total parts in Apollo and its rocket enroute to the moon, giving the U.S. a huge lead over the world in this burgeoning field. Engineers had to design pumps, valves, filters and switches that had to function with unprecedented reliability. They developed miniature medical sensors to monitor astronauts a quarter million miles away; they vastly improved the range of communications equipment; they created compact new electronic parts, long-life power systems, new alloys, new lubricants, new adhesives; they invented ingenious tools and techniques for shaping and joining metals.

Then there was a matter of national prestige. President Johnson argued that the nation, through John F. Kennedy, had made a commitment that should be fulfilled, lest we forfeit space to the Soviets.

The defenders won the debate. But as NASA moved ahead, there were many who wondered if perhaps the launch pad fire had cost America the chance to be first on the moon. The Soviets had not launched a man into space in nearly two years, but they had flown unmanned tests of their next generation spacecraft, the Soyuz, which intelligence reports indicated would be their moon ship.

Soyuz flew manned for the first time on April 23, 1967, 86 days after the Apollo fire, with cosmonaut Vladimir Komarov at the controls.

Soyuz, translated, meant union, and NASA officials expected a second spacecraft to be launched for a linkup with Komarov. But there was no such launch. Periodic flight bulletins reported the cosmonaut was well and the flight was proceeding smoothly.

Then, after 27 hours, the bulletins ceased. For 11 hours there was silence from Moscow, then a terse an-

The five F-1 engines of the huge Apollo/Saturn V space vehicle's first stage leave a gigantic trail of flame in the sky above the Kennedy Space Center seconds after liftoff. The unmanned *Apollo 6* was launched at 7 a.m. April 4, 1968, and was recovered (right) by U.S. Navy frogmen from the U.S.S. *Okinawa* following its 4:58 p.m. splashdown in the Pacific north of Honolulu, Hawaii.

nouncement. The parachutes failed on descent to Earth and Komarov was killed when *Soyuz 1* crashed at a speed of 270 miles an hour in steppes near the town of Orenburg in the Ural Mountains. For the first time a human being had died in space flight.

The Soviet space program, like the American, entered a period of deep questioning, of rebuilding.

While the Apollo command and lunar modules were being redesigned, NASA's unmanned rockets were busy propelling a series of probes to the moon to find out what it was like before humans went there. The surface properties of the moon were largely unknown, and some scientists speculated a spaceship might be swallowed in a deep layer of dust, that electrostatic dust might leap up and destroy a craft's electronics, or that astronauts might fall into thinly-covered crevasses.

These questions had to be answered: How much weight will the lunar surface support? How steep are the slopes? Are there many boulders, and what size? Will the dust or dirt cling? Do the craters pose a threat? What is the precise size and shape of the moon? How strong is the gravity field into which the lunar ship will descend?

The robots — named Ranger, Surveyor and Lunar Orbiter — circled and landed on the moon, dug into its surface, surveyed potential astronaut landing sites, and proved conclusively the moon was a safe place to visit.

There was more good news for NASA as 1967 drew to a close. Wernher von Braun's massive Saturn 5 rocket, 36 stories tall, with power in the first stage alone greater than that of 500 jet fighter planes, had a spectacularly successful debut in November, drilling into orbit an unmanned Apollo craft, a dummy lunar module, and other attached equipment weighing a phenomenal 280,036 pounds — more than the combined weight of all the more than 350 satellites launched previously by the United States.

Two months later, the lunar module, the craft that would ferry two men to the lunar surface, was successfully tested unmanned in Earth orbit.

These were major milestones, and NASA officials, depressed just months earlier, began to feel a manned lunar landing by the end of the decade was still possible. If — and there were many ifs.

This picture of the *Surveyor I* spacecraft shows landing foot No. 2 resting on the lunar surface. Dark area just above foot is depression caused by pressure of the foot as it landed on the moon. The bright spot at bottom is a reflection of the sun.

While work progressed, the agency named the crew that would fly the first redesigned Apollo. Commanding the mission would be Mercury veteran Wally Schirra. Training with him were two rookies, Walt Cunningham and Donn Eisele.

After months and months of redesign and testing, officials set the launch date for the flight, designated *Apollo 7*, for October 11, 1968. They were about to learn if they had profited from the mistakes of *Apollo 1*.

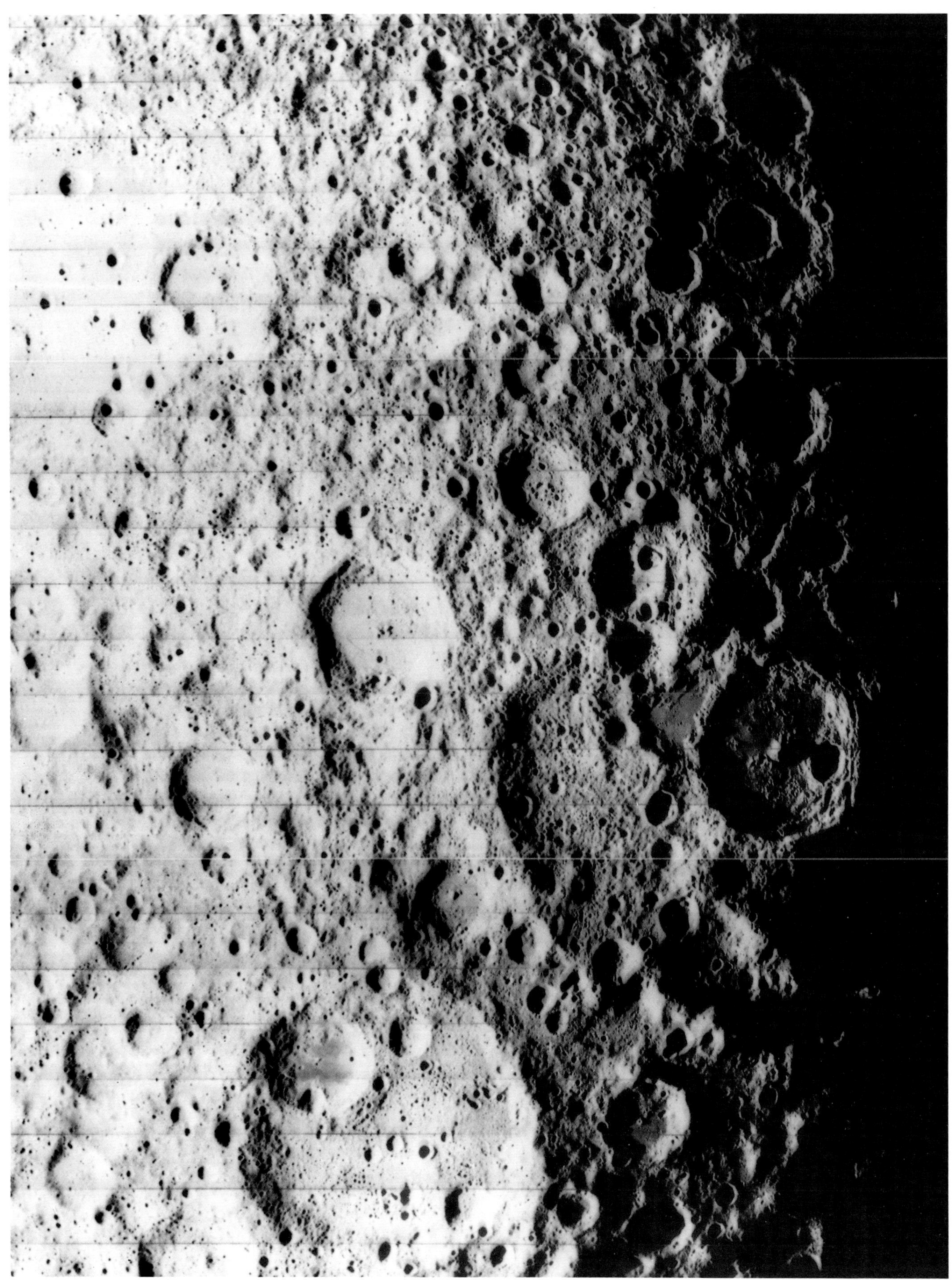

The *Apollo 8* astronauts found the moon to be "a vast, lonely, forbidding place," just as this picture of the moon's backside taken in August 1966 by Lunar Orbiter I predicted. The area of the moon's surface shown is about 810 by 900 miles.

Chapter Eight

'A Vast, Lonely, Forbidding Place'

Long before the October 11 launch, the redesigned *Apollo 7* spacecraft and its Saturn 1B rocket were transported to Launch Complex 34 — the scene of the fire — for weeks of meticulous preparation.

Great care was taken to make certain nothing again would go wrong. The number of people permitted to enter the spacecraft cabin was limited, and they all wore protective clothing and caps. A guard at the hatch checked every nut, bolt and tool that went in and out. Television cameras kept watch on all spacecraft activities.

Women working on the vehicle had to remove cosmetics and rings. Men had to shower to remove loose hair after haircuts.

On the appointed day, Wally Schirra, Walt Cunningham and Donn Eisele were strapped in their couches in what had become known around the Cape as "Wally's ship." Outside, the weather was clear, hot, breezy. In the concrete blockhouse 1,000 feet away, memories sharpened the tension as the countdown advanced toward zero for America's first manned venture into space in nearly two years.

Many of the 300-person launch team had been in these same positions, performing the same duties, on that January day when that chilling cry of "Fire in the cockpit!" had shocked them. Nerves were tight, faces grave, movements calculated and deliberate. Had some little flaw been overlooked? Would more lives be imperiled?

Liftoff was perfect as the Saturn 1B drilled *Apollo 7* into orbit. "Riding like a dream," Commander Schirra radioed.

The *Apollo 7* crew included (left to right) Don F. Eisele, command module pilot; Walter H. Schirra Jr., commander; and Walter Cunningham, lunar module pilot.

Apollo 7 liftoff, October 11, 1968.

For 11 days, the three astronauts tested the spacecraft's systems, conducted experiments, beamed the first extensive live television scenes from a manned orbiting vehicle to fascinated viewers on Earth, and flew their ship longer than would be required for a trip to the moon and back.

They were impressed with the size of the Apollo relative to the cramped cabins of Mercury and Gemini, which had confined the astronauts to their seats. The *Apollo 7* crew members could unstrap themselves and move around the cabin and into a closet-size area beneath the seats, which served as sleeping quarters.

The flight encountered only minor problems, and they were swiftly resolved.

The biggest problem Mission Control had was with the crew. All three had nasty colds and were orbiting the world with stuffy noses. At times, the astronauts were restless and irritable, often declining to carry out added engineering tests that ground controllers devised to gather additional, bonus performance data on a spacecraft that was working so well.

By the final day, the astronauts were in better spirits and thanked ground controllers for "a magnificent job." Back on Earth, Schirra apologized "to everyone for some of my remarks as they were phrased, but not for my decisions."

So successful was the flight that NASA officials boldly considered sending the next manned Apollo

The crew of the *Apollo 8* lunar orbit mission included (left to right) James A. Lovell, Jr., command module pilot; William A. Anders, lunar module pilot; and Frank Borman, commander. They are standing beside the Apollo Mission Simulator at the Kennedy Space Center.

into orbit around the moon to rehearse the techniques needed for getting to and from the lunar vicinity. A rejuvenated Soviet program helped forge that decision.

A month before *Apollo 7*, an unmanned Soviet Zond spacecraft became the first to make a single loop around the moon's backside and return intact to Earth. A second Zond carrying turtles, flies and worms repeated the journey in November.

And on October 26, cosmonaut Georgy Beregovy twice guided his *Soyuz 3* vehicle to within a few feet of an unmanned spacecraft. When NASA officials learned that Zond was a modified version of Soyuz, some were convinced the aim of the Soviets was to upstage America by hurling a cosmonaut on a looping flight around the moon before astronauts could land there.

"It is my belief," said Chris Kraft, the agency's director of flight operations, "that in the following spring they intended to fly a Zond with a man aboard out around the moon and bring him back. That really would have taken the glory, so to speak, out of what we were trying to do with Apollo. Even though they did not land, they could say they were the first to send a man to the vicinity of the moon, that they had beaten the United States.

"What they didn't know was that we were thinking about sending *Apollo 8* on such a circumlunar flight, or even orbiting it around the moon in December. We kept that very close to the vest and very secret."

There was considerable debate within NASA on the wisdom of sending the first men to ride the massive Saturn 5 rocket all the way to the moon. But when the *Apollo 7* results were analyzed and all the engineering tests evaluated, Thomas O. Paine, who replaced the retired Jim Webb as agency administrator, announced in November that the *Apollo 8* astronauts had been assigned to orbit the moon in December.

It was the single greatest gamble in space flight. Some considered it a necessary gamble. Only a year remained before John F. Kennedy's end-of-the-decade goal. Much more information was needed about navigating to and from the moon. So, despite the approaching holidays, the mighty Saturn 5 rocket heaved off its launch pad on December 21, 1968, hurling astronauts Frank Borman, Jim Lovell and Bill Anders away from Earth at 24,200 miles an hour on man's riskiest and deepest penetration into the void of his universe.

The target was the moon — 239,000 miles away.

Three days later, millions on Earth watched transfixed at the spectacular scene on their television screens, a gleaming marble-like sphere. It was a picture that gave humans a new perspective of their planet, for it was a snapshot of one whole face of the Earth as never seen before, taken from a manned spaceship more than 200,000 miles away.

Part of the globe was in shadow, part in sunlight, with great patches of clouds. Because the picture was black and white, Jim Lovell served as tour guide: "For colors, the waters are all sort of royal blue; clouds are bright white. The land areas are generally brownish to light brown.

"What I keep imagining," Lovell mused, "is if I'm some lonely traveler from another planet what I'd think of the Earth at this altitude — whether I'd think it was inhabited or not."

"You don't see anybody waving, do you?" quipped Mission Control.

Those television pictures of Earth had an overwhelming and lasting impact. People began to realize that their planet was indeed a fragile entity floating in a vast universe, a spaceship in its own right, needing attention to its habitability. They felt like never before that Earth was a beautiful planet, worth working to save.

Minutes after the astronauts signed off the television

The *Apollo 8* flight controllers inside Mission Control.

This *Apollo 8* telephoto shows the rising Earth about five degrees above the lunar horizon, which is about 350 statute miles from the spacecraft. On the Earth 240,000 statute miles away, the sunset terminator crosses Africa. The South Pole is in the white area near the left end of the terminator. North and South America are under the clouds.

show, *Apollo 8* reached the "equigravisphere," a point where the gravitational pull of Earth and moon are equal. Seconds later, less than 40,000 miles from the moon, Borman, Lovell and Anders came under the prime influence of lunar gravity. During the long uphill coast from Earth they had fought against Earth's gravity and their speed had slowed to only 2,200 miles an hour. Now they were being pulled toward the moon and their speed gradually increased.

Early on Christmas Eve, when the astronauts received the go-ahead to fire into lunar orbit, they were traveling 5,800 mph, fast enough to swing the ship back toward Earth if a braking rocket failed to slow the craft to orbital speed.

Apollo 8 slipped around the moon's backside, where the crucial firing would take place, out of radio contact with ground control. Not until the spacecraft emerged from behind the moon did controllers know it was in orbit. "We've got it! We've got it!" shouted an excited capsule communicator. "*Apollo 8* is in lunar orbit."

Lovell, the first to describe the moon from only 70 miles away, described it as "essentially gray, no color. It looks like plaster of paris, a sort of grayish beach sand." The first of many questions about that alien world had been answered. During the next 20 hours would come answers to many more.

Throughout the day, the astronauts televised pictures both wild and wondrous. Viewers were awed by scenes of a bleak, barren landscape of crater-pocked plains, boulder-strewn plateaus and rugged highlands.

"It looks like a vast, lonely, forbidding place, an expanse of nothing," Borman said.

The moonmen surveyed and photographed all five of the prime sites being considered for Apollo landing missions. They identified and tracked lunar landmarks and discovered new ones.

By evening, the astronauts reported they were "very tired" but so exhilarated by their achievements, they expanded a planned 15-minute telecast to 37 minutes. While their camera scanned the lifeless lunar landscape, blinding in the sunlight, forbidding and secret in the shadow, and with millions on Earth watching — some already celebrating Christmas, others just hours away, Borman, a lay reader in his church, offered a prayer from space.

A brief pause, and then the three took turns reading the story of the creation, from the first 10 verses of the Book of Genesis. "In the beginning, God created the heaven and the Earth...," Anders began.

Borman concluded the reading, and then sent the world a special Christmas message: "And from the crew of *Apollo 8* we close with a good night, good luck, a Merry Christmas, and God bless all of you — all of you on the good Earth."

Early on Christmas Day, the spacecraft sped behind the moon for the tenth and final time. Again, the big engine fired perfectly. "Please be informed there is a Santa Claus," Lovell radioed as the ship reappeared. "The burn was good."

Apollo 8 was on the way home, "right down the corridor and on target."

The return trip was flawless and early on December 27, the ship slammed into the outer boundary of Earth's atmosphere, 400,000 feet high, at 24,630 mph, 7,000 mph faster than any previous manned reentry. Three large parachutes unfurled at 10,000 feet and *Apollo 8* floated gently to a landing in the Pacific Ocean, just three miles from the recovery carrier. The craft splashed down, appropriately, west of Christmas Island.

Congratulations poured in from around the world. Soviet President Nikolai Podgorny called it "a new achievement in man's mastering cosmic space." Hard words for a Soviet to say. For *Apollo 8* essentially eliminated the Russians from the moon race, their plan to upstage the Americans with a Zond mission around the moon completely overshadowed by the brilliant flight of Borman, Lovell and Anders. The Soviets quickly abandoned that idea.

It was an uplifting event that restored a sense of pride in the United States after a year of turmoil that had recorded the assassinations of Robert Kennedy and Martin Luther King, riots in major cities and at the Democratic National Convention in Chicago, violent protests against the war in Vietnam and, in the furor over the war, Lyndon Johnson's decision not to seek reelection.

Apollo 8 was selected the No. 1 story of the year by newspaper editors. A friend telegraphed Frank Borman: "You have bailed out 1968."

One major piece of Apollo hardware needed to be tested by man in space. That was the lunar module, a craft with four spindly landing legs that looked like a creature from another world — hardly a spaceship intended to land two men on the moon.

Assigned the task of flying the LM, as it was called, in Earth orbit was the *Apollo 9* crew of space veterans Jim McDivitt and Dave Scott and rookie Rusty Schweickart. Because two spacecraft would be involved in the exercise, the crew gave each a code name. The LM was called *Spider* and the Apollo command module *Gumdrop*, after their appearances.

A Saturn 5 thrust the combination into space on March 3, 1969. Once in orbit, the astronauts separated *Gumdrop* from the third stage, turned the command module around and linked it nose-to-nose with *Spider*, pulling it free from its garage-like compartment atop the rocket stage. The third stage then was sent a safe distance away by a ground-commanded radio signal.

Five days after reaching orbit and after checking all LM systems, McDivitt and Schweickart crawled through a 47-inch connecting tunnel and secured themselves in the moon taxi, sealing themselves off from Scott, who remained with the command module.

An excellent view of the docked *Apollo 9* command and service modules and lunar module, with Earth in the background, during Scott's stand-up EVA activity on the fourth day of the *Apollo 9* mission.

Spider, the *Apollo 9* lunar module, undergoes testing by McDivitt and Schweickart on the fifth day of the *Apollo 9* Earth-orbital mission. *Spider's* landing gear is deployed.

They "cast off" from *Gumdrop* and became the first astronauts ever to fly aboard a spacecraft designed to operate only in the vacuum of space. Any lunar module fired from Earth was destined never to return. It could descend to a world without atmosphere like the moon, but would burn up racing into the atmosphere of the Earth. Astronauts flying this machine had to make it back to the mother ship if they wanted to return safely to Earth.

McDivitt and Schweickart fired the descent engine intended to land the craft on the moon and moved into a different orbital path and 113 miles away from Scott. At this point they broke *Spider* in half. The bottom half, holding the descent engine and landing legs, fell away, leaving the two astronauts in a legless, seemingly helpless ascent stage, which bore the engine designed to blast explorers off the moon and carry them to a rendezvous with the lunar orbiting command module. On liftoff from the moon, the descent stage would serve as a launching pad.

McDivitt and Scott triggered the ascent stage and executed a complex series of maneuvers that brought them back to *Gumdrop*. As Scott moved in for a linkup, he commented, "You're the biggest, friendliest, funniest-looking spider I've ever seen." They docked after a separation of more than six hours.

Apollo 9 returned to Earth after 10 days aloft. Only one more flight remained before Americans would be ready to challenge the moon's surface.

Aerial view of Launch Complex 39 at Kennedy Space Center, showing *Apollo 10* space vehicle on its way to Pad B atop a huge crawler-transporter.

The *Apollo 9* crew, (left to right) Russell L. Schweickart, David R. Scott, and James A. McDivitt, walk from the recovery helicopter aboard the U.S.S. Guadalcanal. Their March 13, 1969 splashdown concluded a successful 10-day mission.

Two months later, on May 21, *Charlie Brown* and *Snoopy*, the spaceships of *Apollo 10*, eased into lunar orbit, this time on a mission not only to test the LM further, but also to perfect navigating around the moon and verifying a landing site.

Scouting the Sea of Tranquility, so named by ancient astronomers who thought it was a smooth body of water, was their main objective. If one particularly level plain of that "sea" proved acceptable, then the astronauts of *Apollo 11* soon would be heading for that same area.

Drawing the assignment for this final dress rehearsal, the flight to tie up all the loose ends, were three space veterans: Tom Stafford, Gene Cernan and John Young.

Stafford and Cernan were to fly *Snoopy* to within nine miles of the Sea of Tranquility. Some critics asked, "If they're going that close, why not land?" NASA's reply: Too many unknowns. To risk a landing after only one flight around the moon with a LM that had been through a single manned flight test would be "premature and foolhardy."

The *Apollo 10* crew included (left to right) Eugene A. Cernan, lunar module pilot; John W. Young, command module pilot; and Thomas P. Stafford, commander.

Mission Control, on the second day of the *Apollo 10* lunar orbit mission, receiving a color television transmission. The spacecraft was about half-way to the moon, or approximately 112,000 nautical miles from Earth, when this picture was made.

As Stafford and Cernan swooped toward the Sea of Tranquility, Stafford exclaimed, "There are enough boulders around here to fill up Galveston Bay. It's a fascinating sight." Cernan's voice, too, rang with excitement: "We're right there! We're right over it! I'm telling you, we are low, we are close, babe!"

"All you have to do is put your tail wheel down and you're there!" shouted Stafford.

Snoopy whipped low over the moon, actually four miles south of the intended *Apollo 11* landing site because of navigational errors that planners had expected and intended to learn from.

After a second pass over the Sea of Tranquility, the two astronauts prepared for the critical dismembering of the LM — separating the lunar craft so the legless upper section would return them to *Charlie Brown*.

"Son of a bitch!" The curse from Cernan sent a shock wave through Mission Control.

"We've got some wild gyrations," Stafford reported. *Snoopy* was pitching up and down, yawing left and right.

"Hit the AGS!" Cernan yelled to Stafford to activate the abort guidance system. Stafford's hands flew across his controls. He hit the switch and the LM calmed down.

"I don't know what the hell that was, babe," a relieved Cernan radioed. "But that was something. We were wobbling all over the sky."

Had the two astronauts not reacted as quickly as they did, in another two seconds, *Snoopy* would have been locked into a disastrous dive toward the moon which they could not have survived.

It was too close. Engineers later traced the problem to an abort system switch that had somehow snapped into an incorrect position. That had sent the LM into a radar search for the mother ship, *Charlie Brown*, instead of stabilizing itself for separating its upper from its lower stage.

Finally separated, Stafford and Cernan fired the ascent stage engine and flew their vehicle back to the command module, orbiting 60 miles above the surface. The ships docked, and a happy Stafford reported, "*Snoopy* and *Charlie Brown* is hugging each other. We is back home — almost."

Before starting the homeward journey, the astronauts jettisoned the ascent stage into its own useless orbit and made one more trip around the cratered moon. As John Young took a final look at this rugged world, he remarked, prophetically: "You've often heard the nursery rhyme about the man in the moon. We didn't see one here, but pretty soon there will be two men on the moon."

Chapter Nine

'One Giant Leap for Mankind'

The preliminaries finished, all attention focused on the main event. It was time for *Apollo 11* and the journey of the ages.

NASA engineers consulted their computers for a launch date. The moon is in position only four or five days a month to receive visitors, and added to that restriction was a requirement for a sun angle of between 6 and 20 degrees at the Sea of Tranquility, making it easier for the astronauts to sight shadows cast by lunar landmarks.

The computer cranked out a launch time of 9:32 a.m. EDT Wednesday July 16, 1969, with touchdown on the moon four days later.

Throughout the history of exploration there never had been an expedition planned and prepared with such care as this one. Twenty U.S. man-in-space missions and dozens of unmanned flights preceded the journey, probing all the uncertainties, perfecting the procedures. More than 400,000 government, university and industrial personnel backed up the astronauts, designing and building the hardware, developing the guidance equations.

Unlike Columbus, the three astronauts — Neil Armstrong, Buzz Aldrin and Mike Collins — knew where they were going. They spent hundreds of hours rehearsing the trip, in flight simulators, helicopters and a device that simulated the moon's gravity, one-sixth that of Earth's. Armstrong and Aldrin concentrated on the lunar module they would ride to the lunar surface, while Collins mastered the command module, in which he would orbit the moon as a "lonely lifeguard" while his crewmates were on the Sea of Tranquility.

Armstrong, the commander, and Aldrin also spent hours with geologists learning about the type of rocks they should look for, collect and bring home. For they were going not just for the ride, but to learn some of the secrets of this ancient and alien body. What they learned could unlock some mysteries about the origin of our solar system.

How did these three men draw this assignment? It was the luck of the draw.

Crew selections were made by Deke Slayton, chief of flight crew operations, and Alan Shepard, chief of the Astronaut Office, both of whom had been

In preparation for their moon landing flight, Neil Armstrong, left, trains in the Apollo Lunar Module Mission Simulator in the Kennedy Space Center's Flight Crew Training Building at Cape Canaveral, while Michael Collins, right, trains in the Apollo Command Module mockup in Building 5 of the Manned Spacecraft Center in Houston.

The *Apollo 11* crew leaves Kennedy Space Center's Manned Spacecraft Operations Building for the van ride to Launch Complex 39A during the pre-launch countdown.

grounded by medical problems. In 1968, they named five crews for 1969 missions, hoping that one of the final three trips would achieve a lunar landing.

Jim McDivitt's crew drew *Apollo 9*, testing the lunar module in Earth orbit; Tom Stafford's *Apollo 10* team orbited the moon with a LM. If either of those flights failed, the next crew in line would repeat it. Most space agency observers were betting early in the year that Pete Conrad's *Apollo 12* would probably draw the first landing assignment.

But *9* and *10* were both wildly successful, opening the door and the pages of history for Armstrong, Aldrin and Collins.

Two nights before the launching, the three of them held a news conference, with reporters asking questions via television from a building 15 miles away because the astronauts were in a pre-launch quarantine intended to prevent them from catching a cold or other disease. Said Armstrong, "After a decade of planning and hard work, we're ready and willing to attempt to achieve our national goal."

On launch day, technicians pumped 525,000 gallons of fuel into the Saturn 5 as the countdown progressed smoothly. More than one million people gathered to watch the historic launching, crowding river banks, highways and other viewing areas. Assembled at a VIP site at the Kennedy Space Center were congressmen, diplomats and movie stars.

It was time.

Great clouds of burning fuel, orange in the gray-black smoke, billowed from the base of the rocket, almost right on the designated millisecond, at 9:32 a.m. EDT. The onlookers cheered, prayed, cried or looked on with disbelief as the mighty Saturn 5 punished the launch pad with its dazzling sheet of flame and then thundered like some giant wailing banshee into the blue sky, arcing out over the Atlantic Ocean.

An estimated 500 million people watched on television as three brave men from planet Earth departed for the moon, 250,801 miles away.

"It certainly looks as if you're on your way," Mission Control Center in Houston radioed the trio.

Early on Saturday, the astronauts raced into the shadow of the moon and reported a spectacular first sighting of their target — against a backdrop of a brilliant solar corona as the moon eclipsed all but the halo of gases surrounding the sun. "It looks like an eerie sight," commented Armstrong.

The spaceship slipped behind the moon, where Collins fired the main engine to settle *Apollo 11* into an orbit just 62 miles above the lunar surface.

The crew relayed vivid color television pictures of the rugged landscape, prompting Armstrong to remark, "The view of the moon is really spectacular."

As *Apollo 11* zipped over the cratered Sea of Tranquility plain where Armstrong and Aldrin were to land the next day, the commander said, "We're getting our first view of the landing site approach. The pictures and snapshots brought back by *Apollos 8* and *10* have given us a pretty good preview of what to look at here. It looks very much like the pictures, but it's like the difference between watching a real football game and one on TV. There's no substitute for actually being there."

None of the astronauts slept very long before awakening on landing day. Collins logged six hours and Armstrong and Aldrin five hours each. Mission Control alerted them at 7:02 a.m. They ate breakfast, and Armstrong and Aldrin donned their moonwalking suits, then transferred through a tunnel into the lunar module.

Coasting toward the moon, *Apollo 11* astronauts captured this picture of the solar corona. The moon is the dark disc between the spacecraft and the sun.

"The *Eagle* has wings!" Armstrong reported as the two ships, now separated, emerged from the far side of the moon. The *Eagle* is in lunar landing configuration.

As the world waited for the incredible moment, prayers for a safe journey were offered in many churches, in many lands, on this Sunday morning in July.

Armstrong stood at the commander's post at the left of the LM cabin, Aldrin at his right. Both were restrained by harnesses. They extended the four landing legs and reported they were ready. Two minutes before *Apollo 11* disappeared behind the moon on its 13th orbit, Mission Control relayed the word they wanted to hear: "You're go for undocking."

There were tense minutes on the ground until the two ships, now separated, emerged from the far side and Armstrong gleefully reported, "The *Eagle* has wings!" Once apart, the ships went by radio code names selected by the crew — *Eagle* for the LM, *Columbia* for the command ship.

Collins fired his maneuvering jets to move a safe distance away. "See you later," he said. *Eagle* was on its own.

Behind the moon again, on the 14th orbit, *Eagle's* descent engine was fired, slowing the LM and dropping it into an orbit with a low point just 9.8 miles above the surface. Once they reached this nearest approach, Armstrong had five seconds to decide whether to fire the descent engine again to drop to the surface, or to make another orbital turn before making the burn.

Astronauts and ground controllers monitored the lunar craft's systems. In the control center, capsule communicator Charlie Duke radioed, "You're go for powered descent."

Armstrong pressed a button, the engine started and flames gushed beneath the spindly craft. Landing on the Sea of Tranquility was 12 minutes away.

The two astronauts swooped closer, skimming dangerously low over the moon's hostile mountains, seeking landmarks that would guide them to their target.

Suddenly, computer alarm lights flashed in the cabin. The computer was overtaxed and they were rapidly approaching an abort situation, a hellish maneuver where they would explosively separate the ascent stage and push it to the limits in flying back to the safety of the command ship.

Quick and cool thinking by the astronauts and ground controllers saved the day. A young guidance officer in Mission Control, 27-year-old Stephen Bales, understood the LM computers better than anyone, had worked with them, programmed them. He saw that *Eagle's* computer was doing all the right things. It was just being asked to display too much landing information on the state of every system in the vehicle.

Through Charlie Duke, he told the astronauts to stop asking the computer for so much information, and Armstrong and Aldrin were told to press on toward the surface in a series of dramatic "go's."

"You're looking great, *Eagle*; you're go for landing," Duke called up.

"Roger, understand, go for landing," Aldrin responded. "3,000 feet...2,000 feet...Okay, it looks like it's holding."

Armstrong and Aldrin were jolted as they neared touchdown. The automatic landing system was taking them into an area strewn with large boulders and pocked with craters. Armstrong quickly took manual control and skimmed over the littered field, like a pilot flying a helicopter.

The commander needed 90 seconds to find a smooth spot four miles away. That consumed valuable fuel. He had only 16 seconds of fuel left in the tank when he shut down the engine and settled onto a relatively flat field lying between some sizable craters and another boulder field.

Five-foot-long probes that dangled like curb feelers from three of the legs touched the surface, igniting a light in the cabin. *Eagle* touched down gently, coming to rest at a slight angle.

Dr. Robert R. Gilruth, MSC director, (left photo) and George M. Low, manager of the Apollo Spacecraft Program (in foreground of right photo) watch television monitors in the Mission Control Operations Room during the *Apollo 11* lunar landing mission.

"Houston, Tranquility Base here. The *Eagle* has landed," came the momentous words from Armstrong.

Man landed on the moon at 4:17.42 p.m., EDT, on Sunday, July 20, 1969.

"Roger, Tranquility," replied capsule communicator Duke. "We copy you on the ground. You got a bunch of guys about to turn blue. We're breathing again. There are lots of smiling faces in the room and all over the world."

"There are two of them up here," Armstrong advised.

"And don't forget the one in the command module," chimed in Collins from his lonely orbit in *Columbia*.

Armstrong and Aldrin thoroughly checked the LM systems for three hours, making ready for a quick takeoff in case of an emergency. With no problems, and excitement running high, they asked permission to step outside early, skipping a four-hour rest period. Mission Control concurred.

While waiting to go out, Aldrin, a deeply religious man, privately celebrated communion with bread and wine.

The two astronauts pulled on their boots, gloves, helmets and backpacks. They depressurized the cabin, and Armstrong opened the hatch. He climbed out backwards, cautiously stepped onto a platform and backed down a nine-rung ladder mounted on one of the landing legs.

On the second step, he pulled a lanyard which opened a compartment, uncovering a television camera that permitted the world, in grainy black and white pictures, to share a great moment in history.

An estimated 528 million people, the largest TV audience ever, watched breathlessly as Armstrong backed down the remaining steps, resembling a ghostly figure as he dropped to a strange new world. He placed both feet in a dish-shaped footpad at the bottom of the ladder.

Tentatively, unsure of what was below him, he extended his left foot, encased in a size 9 1/2 boot, and pressed it into the dust of the moon.

Man first touched the moon at 10:56.20 p.m. EDT, Sunday, July 20, 1969.

"That's one small step for a man; one giant leap for mankind," said Armstrong, speaking words that would be forever etched in history.

With an almost shuffling gait, Armstrong began to move about in the harsh light of the lunar morning. His first steps were tentative tests of the moon's soil and his ability to operate in a gravity field with only one-sixth the pull of that on Earth. In his bulky space suit he would weigh 360 pounds on Earth. On the moon, he weighed 60 pounds.

"The surface is fine and powdery," he reported. "It adheres in fine layers, like powdered charcoal, to the

Buzz Aldrin leaves the lunar module to join Neil Armstrong on the moon's surface.

Neil Armstrong (left) and Buzz Aldrin deploy the United States flag on the moon. This photo was taken by a 16mm data acquisition camera mounted on the lunar module.

soles and sides of my boots. I only go in a fraction of an inch, maybe an eighth of an inch."

Armstrong moved a few feet from the ladder and extended a "collection sample rod" to gather about two pounds of soil which he put in a plastic bag and stuffed in a suit pocket. The astronauts planned to collect several pounds of rock and soil during about two hours outside the LM, but if they had to make a sudden emergency departure from the moon, this small initial sample would prove invaluable for research scientists back on Earth.

Then the first human being on the moon permitted himself a long look at the level, rock-strewn moonscape around him. "It has a very stark beauty all its own," he described. "It's like much of the high desert areas of the United States. It's different, but it's pretty out here."

Aldrin stepped onto the surface 18 minutes after Armstrong's first step.

"Beautiful! Beautiful! Magnificent desolation," he exclaimed.

He was struck with the shocking contrasts of color. There were many shades of gray, a pale tan, and areas of utter black where rocks cast their shadows along the airless surface.

In the distance, the Earth was a resplendent oasis of shifting colors, and many times brighter in places where sunlight splashed off the clouds and oceans.

The astronauts carried a television camera about 60 feet away and mounted it on a post so that spellbound Earth viewers could watch the dramatic moon expedition. The pictures were grainy, but fascinating. The centerpiece was *Eagle*, resting on its four spindly legs among the rocks and silhouetted vividly against the deep blackness of space.

Earthlings watched as the astronauts continually tested their ability to move in the low gravity field, loping like antelopes and bouncing like kangaroos. "It's not difficult at all," Armstrong noted.

Armstrong and Aldrin planted a 3-by-5-foot American flag and Aldrin stepped back and saluted it. Wire stiffeners enabled it to stand out to its full width, since otherwise in the airless vacuum it would have draped on its pole.

But they made it clear they came as ambassadors of all mankind when they unsheathed a stainless steel plaque left on the moon. As Commander Armstrong read the plaque's words, his voice carried throughout the world. "Here men from planet Earth first set foot upon the moon, July 1969. We came in peace for all mankind."

Plaque left on the moon reads: "Here men from planet Earth first set foot upon the moon, July 1969. We came in peace for all mankind."

Buzz Aldrin carries two components of the Early Apollo Scientific Experiments Package (left photo) which he has deployed on the moon's surface (right photo).

They deposited on the moon a disc inscribed with messages from the leaders of 73 nations and they carried to the surface and brought back to Earth the flags of 136 nations.

In a radio link from the Oval Office, President Richard M. Nixon told Armstrong and Aldrin: "Because of what you have done, the heavens have become part of man's world. And as you talk to us from the Sea of tranquility, it inspires us to redouble our efforts to bring peace and tranquility to the world."

Replied Armstrong: "It is a great honor and privilege for us to be here representing not only the United States but men of peace of all nations, men with interests and curiosity and men with a vision of the future."

The explorers deployed a seismometer to radio to Earth, long after they left, information on quakes and rumbles within the moon and to record the strikes of meteorites smashing into the surface. A device to measure the flow of radiation particles from the sun went into position.

They described in detail their wondrous world as they moved about, using their geology training to collect about 50 pounds of lunar material, which they packed in airtight containers for the trip back to Earth.

Mission Control informed them shortly after midnight that it was time to return to *Eagle's* cabin. Aldrin climbed back aboard after 1 hour 44 minutes on the surface. Armstrong scaled the ladder after 2 hours 14 minutes outside.

Ecstatic with elation and excitement, the astronauts slept fitfully for five hours in the cramped LM before starting the countdown to liftoff.

Twenty-one hours after landing, Armstrong fired up the ascent engine, and *Eagle* vaulted free of its launch platform, the lower half of the lunar module.

The firing was perfect, and an exuberant Armstrong announced: "The *Eagle* is back in orbit!"

Mike Collins and *Columbia* orbited 300 miles ahead. For three and a half hours, Armstrong and Aldrin skillfully flew their ship through a series of maneuvers that zeroed in on the command module. As *Eagle* moved in for a linkup, Collins reported it was "steady as a rock," and the two vehicles were once again one.

The moonmen transferred their lunar booty into the command ship, and Collins reported they were "all smiles and giggles over our success." They discarded the faithful *Eagle*, leaving it to orbit the moon for several weeks before lunar gravity dragged it down to a crash landing.

Eagle leaves the moon and heads for a docking rendezvous with *Columbia*. The Earth rises above the moon's horizon.

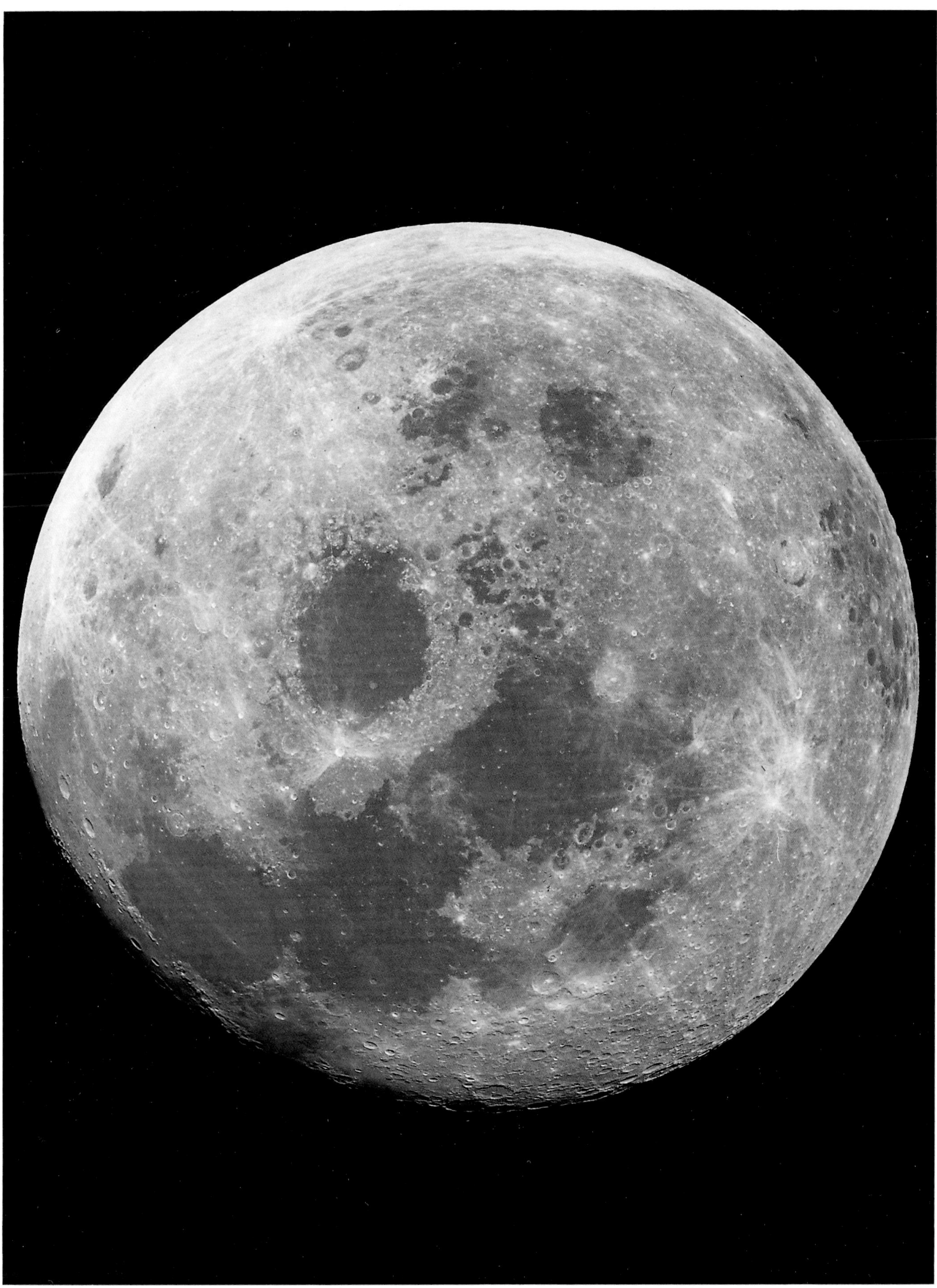

Headed back to Earth, the *Apollo 11* astronauts took this picture of the moon approximately 10,000 nautical miles from the moon.

President Richard M. Nixon welcomes the *Apollo 11* astronauts aboard the U.S.S. *Hornet* recovery ship. The astronauts, confined in the Mobile Quarantine Facility, are (left to right) Neil Armstrong, commander; Michael Collins, command module pilot; and Edwin E. Aldrin, Jr., lunar module pilot.

Wearing special germ-free clothing, Dr. Robert R. Gilruth and NASA scientists look at lunar samples collected by Apollo astronauts.

The astronauts fired up their big engine and began a 60-hour homeward journey that was restful and uneventful. The historic flight ended in the Pacific on July 24, 8 days, 3 hours, 18 minutes after liftoff from Florida. Splashdown was nine miles from the recovery carrier *Hornet*.

Among those on the carrier to greet the new heroes was President Nixon, even though he could not shake their hands. Armstrong, Aldrin and Collins immediately went into a quarantine routine which lasted 16 days, to protect Earth from any possible contamination from moon germs. Nixon spoke to the astronauts through a glass window in an isolation van that they rode all the way back to Houston, by ship, plane and truck.

In Mission Control, controllers laughed, cheered, smoked victory cigars and waved small American flags. Flashed on a large display screen were the 1961 words of President John F. Kennedy when he put the nation on course to the moon. Above it was this cryptic message: "Task accomplished...July, 1969."

Back in Houston after a three-day trip in their van, the astronauts transferred to a Lunar Receiving Lab, where volunteer doctors, technicians, a cook and a few others remained in isolation with them until the quarantine ended. The pilots told of their flight, underwent medical exams and relaxed.

The moon rocks and soil went to another part of the lab, and scientists began the years-long job of examining these precious bits of material from another world. The material was handled carefully in vacuum chambers, with researchers inspecting and cutting it while using glove ports.

The astronauts were pronounced extremely healthy, with no alien germs to report, and they were released to go home to their families.

Once the three were free, their privacy vanished. The United States expressed its gratitude and millions turned out to hail the new heroes during tours that took them to many cities around the world.

Collins left the astronaut corps soon after the flight and joined the U.S. State Department as an assistant secretary for public affairs. He later shifted to the Smithsonian Institution as director of the then-building National Air and Space Museum.

Armstrong and Aldrin stayed with NASA a few months longer before resigning. Armstrong became a professor of aerospace engineering at the University of Cincinnati and Aldrin returned to active duty in the Air Force.

Seated (left to right) behind the glass partition, Aldrin, Collins and Armstrong go through their first post-flight debriefing on Sunday, July 27, 1969. In the foreground are Donald K. Slayton (right) MSC Director of Flight Operations, and Lloyd Reeder, training coordinator. The three astronauts were released from quarantine on August 11, 1969.

Chapter Ten

'Houston, We've Had a Problem'

Ten years later, Buzz Aldrin lamented in an anniversary interview that the promise of *Apollo 11* was never fulfilled. A promise of nations pulling closer together, a promise of an end to strife and wars, a promise of even brighter things in space, with countries cooperating in such things as moon bases and manned flights to Mars.

Instead of beginning a new era, it became the end of a series of flights that had a goal, and that goal — landing on the moon — had been achieved. Time to turn attention to other things.

"I've given a lot of thought in the last couple of years to try to understand the decline of public interest in the space program after *Apollo 11*," Aldrin said. "I think somehow there was a misinformation, in a sense a lie, that we got caught up in and that we disseminated.

"It has to do with why we went to the moon. It was clearly motivated by Sputnik and by our frustrations at our original forays into space. We were badly behind the Russians. There were some political things that were going on that also affected the credibility of the United States in relation to the rest of the world. The developing countries — the ones that had always looked up to us — weren't always looking up at us anymore. Many were looking up at the Soviets because of their space successes. We were very frustrated. We were looking at our education system, at our engineers.

"President Kennedy very rightly read this and decided we had to do something," Aldrin said. "To do what? To chart a course in this new frontier, this new ocean called space. And, clearly, when he said that within this decade we're going to send a man to the moon and bring him back safely, that was a challenge with the full awareness we would do it before the Russians.

"It continued on that way another four or five years, and then some people, especially scientists, began to ask, 'What are we doing this for?' And the answer started coming back — partially to satisfy the scientists — that we were going to learn the origin of the moon, and from that we're going to learn about the origin of the Earth, and we're going to get all these technology spinoffs.

"We never would have sold the program for those reasons in the beginning," Aldrin said. "So when we got there, it didn't carry with it the great sense of national pride and prestige that we had set out to do. It was, look at what we did. We're bringing back all this moon stuff.

Dr. Christopher C. Kraft, Johnson Space Center Director, speaks at the 10th anniversary celebration of *Apollo 11* at the JSC space park. Behind him is a huge, unused Saturn 5 launch vehicle, sister ship to the rocket that sent astronauts to the moon.

Buzz Aldrin, shown here speaking at the 20th anniversary of the *Apollo 11* moon landing, lamented at the 10th anniversary that the promise of *Apollo 11* to pull nations together to cooperate in space projects had not been fulfilled. With the international space station, that promise is becoming reality.

The *Apollo 12* crew included (left to right) Charles Conrad Jr., Richard F. Gordon Jr., and Alan L. Bean. The clipper ship on the *Apollo 12* crew insignia (right photo) signifies an all-Navy crew and symbolically relates the era of the clipper ship to the era of space flight. As the clipper ship brought foreign shores closer to the United States and marked the increased utilization of the seas by this nation, spacecraft have opened the way to other planets. The portion of the moon shown represents the Ocean of Storms area where *Apollo 12* landed.

"I really think that's sad when you can't take more pride in what you have accomplished. One of the reasons is the country got caught up in the confusion of why we went there.

"The real reason was downplayed. It was that we might lose."

Atop a huge crawler-transporter, *Apollo 12*, complete with Saturn 5 stack and its mobile launch tower, rolls out of the Vehicle Assembly Building to Launch Pad 39A.

Indeed, the United States had won the race to the moon and proved technological mastery over the Soviets. The goal accomplished, the impetus was gone. There were other problems to focus on: An American public torn by the war in Vietnam, racial and campus unrest.

A long, troubled period began for NASA. From a peak budget of $6 billion in 1965, at the height of Apollo development, the agency's fortunes had been declining every year, and the fiscal 1970 budget dropped below $4 billion. Tens of thousands of aerospace workers lost their jobs.

NASA's proposals for permanent large space stations, a research base on the moon and manned journeys to Mars were derailed by the Nixon administration and Congress. The only major proposal to survive was a reflyable space plane — the space shuttle. It lived because President Nixon felt continued flights of astronauts into space was good for the morale of the country, and the shuttle left open the option of perhaps one day building a permanent space station.

The budget ax also forced officials to cancel three Apollo flights. But six remained to learn as much as possible about the moon.

Apollo 12's Pete Conrad, Alan Bean and Dick Gordon set sail on November 14, 1969, for the moon's Oceanus Procellarum — Ocean of Storms. With President Nixon among tens of thousand watching, the Saturn 5 rocket had an extremely stormy departure.

Wet observers could not see the rising booster through the rain and clouds. But suddenly they were surprised and alarmed when lightning flashed from the low clouds into the launch area.

The three Navy commanders inside the command module they had named *Yankee Clipper* saw brilliant flashes. The cabin went suddenly dark, then came alive with flashing alarm lights.

"I think we got hit by lightning," commander Conrad radioed, sending a chill through Mission Control.

But backup batteries automatically came on line, and Gordon promptly punched circuit breakers to restore all equipment that had failed. The Saturn 5's separate guidance system had not been affected and continued throughout the crisis to push *Apollo 12* into orbit, where the astronauts made an extensive check to make certain no systems had been damaged.

Three days later they were orbiting the moon, and Conrad and Bean separated the lunar module *Intrepid*. Their goals were to land near an unmanned Surveyor spacecraft that had scouted the moon two years earlier and to start the true exploration of the lunar surface.

"*Apollo 11* proved that man can land on the moon; *Apollo 12* will start the first detailed geological exploration," Conrad explained.

They achieved the first goal when they parked *Intrepid* just 600 feet from the Surveyor. Coming down in the right place was important because pinpoint landings in mountainous areas would be a must for some of the sites being considered for future Apollo landings.

"Outstanding," exalted Conrad. "I can't wait to get outside. Those rocks have been waiting for four and one-half billion years for us to come out and grab them. Holy cow! It's beautiful out there!"

Conrad and Bean performed two four-hour excursions from the lunar module, deploying scientific instruments and collecting 75 pounds of rocks and surface material. The made their way down the slope of a wide crater to Surveyor and clipped five parts from the spacecraft so that analysts on the ground could determine how certain materials are affected by long exposure to the harsh environment on the moon.

Soon after, *Intrepid* blasted off from the Ocean of Storms and rejoined Gordon and *Yankee Clipper* for a safe trip home.

Apollo 12 astronaut Charles Conrad Jr. examines the unmanned Surveyor III spacecraft which soft-landed on the moon on April 19, 1967, paving the way for Apollo landings. The television camera and several other pieces were taken from the Surveyor and brought back to Earth for scientific examination. The lunar module is in the right background.

Apollo 12 lunar module pilot Alan Bean deploys components of the Apollo Lunar Surface Experiments Package on the moon's surface.

John F. Kennedy's goal of landing men on the moon and returning them safely to Earth before the decade was out had been achieved, twice.

The moon-whipped Soviets, meanwhile, were shifting their space program onto a different course. They scored a space triple header by launching *Soyuz 6*, *7* and *8* with a total of seven cosmonauts on consecutive days. The trio maneuvered to within a few hundred feet of one another and then each returned to Earth after five days in space.

Officials said the purpose was to bring three ships close together in orbit in a rehearsal for later assembly of a permanent manned space station.

In a television picture, beamed from 205,000 miles in space, astronaut Jim Lovell, the *Apollo 13* commander, and Fred Haise pulled switches in the lunar module *Aquarius*, inspecting it for a landing two days later to search for ancient rocks in the moon's rugged Fra Mauro highlands. Jack Swigert sat at the controls in the command module *Odyssey*.

America's third moon-landing expedition had gone well since launching from the Kennedy Space Center on April 13, 1970, and Lovell told television viewers, "We're just about to close out our inspection of *Aquarius* and get back for a pleasant evening in *Odyssey*. Good Night."

As Lovell recalled later: "A pleasant evening, indeed! Nine minutes later the roof fell in."

The first word of trouble came from Swigert: "Hey, we've got a problem here."

The pilots had been startled by what sounded like an explosion in *Odyssey*'s service module, which contained the command module's main engine and most of its power, oxygen and environmental systems. The ship began to rock slightly, and red and yellow alarm lights flashed in the cabin.

"Houston, we've had a problem," Lovell repeated. "We've had a main B bus interval (power failure)."

"Okay, stand by *13*, we're looking at it," replied capsule communicator Jack Lousma in Mission Control, as controllers snapped to attention at their consoles.

Posing for this official *Apollo 13* crew portrait in April 1970, command module pilot John L. Swigert Jr., center, points to the third lunar landing mission site on a model of the moon. At left is James A. Lovell Jr., commander, and at right is Fred W. Haise Jr., lunar module pilot.

"We've had a pretty large bang associated with the caution and warning," Haise reported.

In Houston, radio data confirmed the current in one of the two major electrical harnesses had dropped sharply and that the pressure in one of the service module's two oxygen tanks had shot upward, then quickly plunged to zero.

The tank had burst, spewing its 320 pounds of liquid oxygen. Lovell looked out a window and said with apprehension, "We're venting something into space. It's a gas of some sort." The ship began to pitch and roll in reaction to the oxygen gas expulsion. Swigert stabilized it by firing small control jets.

All three astronauts, calling on years of test pilot experience, remained calm as the crisis built. "One of the main electrical circuits is dead," Swigert reported. The explosion of the tank also knocked out two of the three fuel cells which provided electricity to the command module.

Apollo 13 was 49,000 miles from the moon, traveling 3,200 mph, but there no longer was any thought of landing there. Instead, thousands of persons on the ground mounted a massive effort to retrieve the astronauts from a quarter million miles in space.

The one remaining fuel cell was sufficient to get them home if it continued to work. But the oxygen tank that serviced it had been damaged in the explosion and it began to lose pressure. Without the fuel cells, *Odyssey* would have no electrical power except for three short-life batteries that must be saved for reentry through Earth's atmosphere at journey's end.

"We're starting to think about the LM lifeboat," capsule communicator Lousma said.

"Yes, that's something we're thinking about, too," Swigert responded.

One hour 29 minutes after the explosion, Mission Control estimated *Odyssey* would be a dead ship in 15 minutes.

Without hesitation or panic, Lovell and Haise moved through the connecting tunnel into the for-now safe haven of the lunar module. *Aquarius* had been built to land them on the moon, remain there for nearly 40 hours, then carry them back to the mother ship. Now it was to serve as their lifeboat, its independent power and oxygen systems sustaining the astronauts until they approached the Earth three and one-half days later — if they survived.

Swigert remained in the dying command module while his companions powered up the LM. He shut down its systems, plunging the cabin into darkness. Groping his way by flashlight he transferred the precise alignment of *Odyssey's* guidance platform to a similar platform in *Aquarius*.

The lunar craft's cabin would be too cramped for all three men for a long period of time. So they borrowed a 10-foot hose from Haise's moon-walking suit, rigged it to the LM oxygen system and ran it into the command module. Thus, one at a time, they could use the more spacious *Odyssey* as a kitchen, bathroom and bedroom, although a very cold one.

They were safe for the moment. But controllers knew the fight for survival was just beginning. Using computers they calculated the maneuvers to get the astronauts home and how best to conserve precious oxygen, water and power.

Without the command ship's powerful engine, there was no way to turn *Apollo 13* around toward the Earth. As it was, it was on a pre-determined course that would swing it around the back of the moon and

This oblique view of the lunar farside was photographed from the *Apollo 13* spacecraft as the damaged ship swung around the moon and headed back to Earth.

The severely damaged *Apollo 13* service module, photographed after it was jettisoned, shows an entire panel blown away by the oxygen tank explosion. "It's really a mess," Lovell reported.

head it back toward Earth. If there were no course adjustment, however, it would miss Earth by 20,000 miles and become stranded in space.

The adjustment would have to be made by the LM descent engine, much smaller than *Odyssey's* powerplant. Lovell triggered the firing 5 hours 35 minutes after the explosion. It was perfect. Two later ignitions fine-tuned the course.

The spacecraft swept behind the moon, approaching to within 156 miles of the backside, and then looped back toward Earth. Little was said about this celestial body that had been the astronauts' target. As they sped away, Lovell remarked, "Boys, take a good look at the moon. It's going to be a long time before anybody gets up here again."

They carefully nursed their oxygen, water and power, and took turns sleeping fitfully in the command module. But it wasn't easy. They knew they were in a dangerous situation with no margin for error or further serious malfunctions. As Lovell said later, "The trip was marked by discomfort beyond the lack of food and water. Sleep was almost impossible because of the cold. When we turned off the electrical systems we lost our source of heat. We were cold as frogs in a frozen pond. It wasn't simply that the temperature dropped to 38 degrees Fahrenheit — the sight of perspiring walls and wet windows made it seem even colder."

As they neared home, the world watched and waited. Once again, the drama of man struggling against the hostility of space seemed to unite people everywhere in a single hope for the safety of three fellow humans. Prayers were offered in many lands. Several governments offered help. British, French and Russian ships steamed toward possible landing areas.

Early on April 17, five hours before *Apollo 13* was to splash down, Lovell fired the lunar vessel's engine one last time to shift course slightly and aim for a target southeast of Samoa in the Pacific. An hour later, the spacemen discarded the service module using a technique worked out by other astronauts in ground simulators.

When they pulled away from the service section, the three men had a frightening view of the damage caused by the explosion.

"There's one whole side of the spacecraft missing," Lovell explained. "The whole panel is blown out from the base to the engine. It's really a mess."

Swigert returned to *Odyssey* to restore its systems for reentry. The cabin was cold and clammy, the walls, ceiling, floor, wire harnesses and panels covered with droplets of water.

He feared that conditions were the same behind the panels and that turning on the electrical power might create dangerous short circuits. But thanks to procedures radioed from Mission Control and safeguards built into the vehicle after the *Apollo 1* fire in 1967, no electrical arcing occurred.

An hour before they hit the atmosphere, at an altitude of about 15,000 miles, Lovell and Haise joined Swigert and closed the hatches at both ends of the tunnel connecting the two spacecraft. They pressurized the passageway with oxygen and then split the tunnel down the middle with explosive charges, blowing *Aquarius* away like a cork popped from a champagne bottle.

"Farewell, *Aquarius*, we thank you," Mission

Apollo 13 splashed down April 17, 1970, in the South Pacific, about four miles from the U.S.S. *Iwo Jima*. A U.S. Navy helicopter lifts Haise in the Billy Pugh net, while Swigert and Lovell wait their turn with a U.S. Navy underwater demolition team swimmer who has secured the spacecraft.

Crewmen aboard the U.S.S. *Iwo Jima* guide the *Apollo 13* command module atop a dolly on board the ship.

President Nixon presented the *Apollo 13* astronauts the Presidential Medal of Freedom in post-mission ceremonies at Hickam Air Force Base, Hawaii.

Lovell, on board the *Iwo Jima*, reads a newspaper account of the safe recovery of the problem plagued mission.

During his visit to the Manned Spacecraft Center April 18, 1970, President Nixon praised *Apollo 13's* flight directors for bringing the spacecraft safely home. Left to right: Nixon, Glynn S. Lunney, Eugene A. Kranz, Gerald D. Griffin, Milton L. Windler, and Sigurd A. Sjoberg, MSC's director of flight operations. Seated at Nixon's right is NASA Administrator Thomas O. Paine.

Control saluted. "She was a good ship," Lovell added.

The command module, now kept alive by its reentry batteries, plunged into the atmosphere at 24,500 mph 400,000 feet above the Pacific. Fourteen minutes later, millions around the world watched on television as *Odyssey* parachuted into the water just three miles from the main recovery carrier.

Prayers of thanks were offered everywhere for the safe, thrilling and dramatic conclusion of man's most perilous space adventure.

President Nixon flew to Houston to present the Medal of Freedom, the nation's highest civilian award, to the team in Mission Control. He continued on to Hawaii, where he presented Medals of Freedom to Lovell, Haise and Swigert.

"You did not reach the moon," he told them. "But you reached the hearts of millions of people on Earth by what you did."

The three astronauts returned to Houston and personally thanked the flight controllers who engineered their safe return.

Two months later an accident review board concluded the explosion was caused by a "serious oversight" in the design and testing of the oxygen tank.

While the Apollo program stood down for several months for correction of the problem, critics again called for the scraping of the moon landing program. NASA prevailed, but officials knew another flawed flight would doom the program and possibly derail future U.S. manned missions.

Chapter Eleven

'Man Must Explore'

Apollo was ready to fly again nine months later. Carrying the banner for *Apollo 14* was the man who 10 years earlier had energized a faltering U.S. space program when he became the first American to challenge this new frontier.

Alan B. Shepard knew the future of the nation's manned space hopes rode on his mission as he, Ed Mitchell and Stuart Roosa boarded their spaceship on January 31, 1971. Fresh in the memories of everyone was the near-disaster of *Apollo 13.*

Two years after his pioneering flight, Shepard was grounded by an inner ear problem. A new and delicate operation finally corrected it in 1968 and he campaigned hard for a moon flight, finally earning his command.

At 47, he was the oldest man yet named to a space flight.

The three astronauts were to complete the *Apollo 13* assignment: To orbit the moon, with Shepard and Mitchell landing in the rugged highlands of the Fra Mauro Formation to search for ancient lunar material.

They almost didn't make it. On the outward journey, three major problems threatened to abort the mission before Mission Control and the astronauts, working feverishly, overcame them. The command ship *Kitty Hawk* and lunar module *Antares* at first failed to link up after the Saturn 5 propelled them toward the moon because of a docking mechanism flaw. A spurious electrical abort signal in the LM computer almost stopped the landing as Shepard and Mitchell headed for a touchdown. That crisis resolved, they then lost their surface radar, reclaiming it just as they reached the moment when they would have had to abort.

"Whew, that was close," said Mitchell.

Antares, named for one of the stars that guided the ships to the moon, settled in the dust of the boulder-littered highlands. As Shepard stepped outside hours later, he remarked: "It certainly is a stark place here at Fra Mauro. I think it's made all the more stark by the fact the sky is completely black."

The largest rock from Apollo 14 moon samples, basketball size and weighing 8,998 grams, was nicknamed "Big Bertha."

During two outside expeditions, Shepard and Mitchell erected a nuclear-powered science station, collected 92 pounds of rocks and gathered deep-down samples by driving core tubes into the surface.

A steeper-than-expected slope that was covered with huge boulders prevented them from reaching the rim of a crater named Cone, gouged out of the top of a 400-foot ridge, where they hoped to gather samples billions of years old. They clamored to within 30 feet of the rim, but with their hearts and pulses racing, Mission Control told them to forget it.

The rest of the expedition went extremely well, and when *Apollo 14* returned to Earth, the three astronauts had erased much of the disappointment of *Apollo 13* and restored confidence in the space program.

As Alan Shepard steps onto the moon, he lifts his right glove over his helmet visor to shade his eyes from the brilliant sun.

Lunar dust can be seen clinging to the boots and legs of Edgar Mitchell as he moves across the lunar surface.

Salyut 1, the Soviet Union's first space station.

Ever since *Apollo 8* and then *Apollo 11* clinched the moon victory for the United States, the Soviet Union had been saying that its intent all along was not the moon, that its goal was to establish a space station as a permanent home in orbit.

The first model of its first-generation station, *Salyut 1*, was rocketed unmanned into a 135-mile-high orbit on April 19, 1971. It was a 19-ton cylinder, roomy as a house trailer, with four compartments, experiment equipment and working space, a bathroom, a kitchen, two docking ports for Soyuz spacecraft and solar panels to draw electrical power from the sun.

Four days later, *Soyuz 10* vaulted into space and began tracking the station. The three cosmonauts aboard successfully docked with their target. But they disengaged after five hours without entering the station and returned to Earth. The Soviets insisted the purpose of the flight was to test new rendezvous and docking procedures and there was no intention to board the Salyut on this initial visit.

Six weeks later, on June 6, *Soyuz 11* shot into space and hooked up with *Salyut 1*. This time the three cosmonauts, Georgi Dobrovolsky, Vladislav Volkov and Viktor Patseyev, slipped through a tunnel into the station. "There seems to be no end to this station," commander Dobrovolsky commented as he surveyed the spacious interior.

For 23 days, they circled the globe, breaking the old space endurance mark of 17 days. They conducted medical, biological and astronomy experiments, observed the behavior of tadpoles, flies and plants in weightlessness and photographed geological formations on Earth.

The Russian press hailed them as the first "space colonists" and said America's moon flights were a temporary thing, that space stations were the wave of the future, and cosmonauts were leading the way.

A heroes' welcome was being prepared as the trio departed from *Salyut 1* on June 29, ready to steer *Soyuz 11* home. "I am starting the landing procedure," Dobrovolsky reported.

Those were the last words heard from the crew. The retro-rockets fired on schedule and *Soyuz 11* dropped out of orbit and began its automatically-controlled descent. Over and over, the control center messaged the craft. But no reply.

Recovery forces were alerted, and two helicopters were overhead minutes after the spaceship landed. The hatch was opened. All three cosmonauts were hanging lifeless in their seat straps.

James Irwin works at the lunar roving vehicle at the Hadley-Apennine landing site. The lunar module *Falcon* is at left, and the undeployed Laser Ranging Retro Reflector lies atop the LM's Modular Equipment Stowage Assembly. Hadley Delta and the Apennine Front are in the background to the left. St. George crater is about 3 statute miles in the distance behind Irwin's head.

Marching with other Soviet cosmonauts in the funeral procession was astronaut Tom Stafford, representing President Nixon. One day Stafford would orbit the Earth with Russian spacemen.

Investigation revealed the cosmonauts died when a leak developed in the spacecraft during reentry and their cabin lost pressure.

Apollo 15 astronauts left this commemorative plaque on the moon in memory of 14 NASA astronauts and U.S.S.R. cosmonauts who had died. Listed in alphabetical order, they are: Charles A. Bassett II, Pavel I. Belyayev, Roger B. Chaffee, Georgi Dobrovolsky, Theodore C. Freeman, Yuri A. Gagarin, Edward G. Givens Jr., Virgil I. Grissom, Vladimir Komarov, Viktor Patsayev, Elliot M. See Jr., Vladislav Volkov, Edward H. White II, and Clifton C. Williams Jr. The tiny, man-like object represents the figure of a fallen astronaut/cosmonaut.

A plaque honoring Dobrovolsky, Volkov and Patsayev was carried to the moon the following month by the *Apollo 15* crew — Dave Scott, Jim Irwin and Al Worden.

Scott and Irwin placed the plaque on the surface after landing their lunar module *Falcon* in a narrow valley hemmed in on three sides by the 15,000-foot-tall Apennine Mountains and on the fourth by a mile-wide canyon named Hadley Rille.

As Scott stepped into the dust, the seventh human to leave his footprints on the moon, he was struck by the beauty of the lunar dawn. "As I stand here on the wonders of the unknown at Hadley," he said, "I sort of realize there is a fundamental truth to our nature. Man must explore. And this is exploration at its greatest."

Apollo 15 was the first mission to carry a lightweight electric car, a cross between a golf cart and a dune buggy. Battery-powered electric motors drove the vehicle on level ground at 7 mph, but zooming downhill, the moonmen zipped to 11 mph.

"Man, oh, man!" Scott exclaimed as the moon buggy raced over an undulating plain. "This is really a sporty driving course. What a Grand Prix this is!"

Both were strapped in with seat belts. "Hang on," Scott cautioned. "Buckin' bronco!" cried Irwin.

Able to carry heavy loads of tools, cameras, rocks and other equipment and with the ability to travel six miles from the lander, the moon car increased tremendously the area traversed, studied and sampled by the astronauts. The six miles was a safety feature. If the buggy broke down, the men would still have enough power and oxygen in their suits to walk back to *Falcon*.

In three separate excursions over three days, Scott and Irwin drove far and wide over the Hadley-Apennine region, climbing slopes, driving into wide, shallow craters, probing the edge of Hadley Rille. They collected 171 pounds of samples and vividly described the scene for scientists before returning to the safety of their command ship *Endeavour*.

Before leaving lunar orbit, the astronauts ejected a 78-pound satellite carrying instruments to send back data for a year on lunar magnetic, electrical and gravity fields. And, 200,000 miles out, enroute home, Worden took a space walk to retrieve film cassettes from two cameras that had mapped the moon.

A second lunar buggy prowled the moon's surface in April 1972, as astronauts John Young and Charles Duke explored a wide plateau in the Descartes Mountains, the highest topographical region on the moon's front side.

"Look at those boulders! Look at those rocks! They get bigger and bigger!" Duke explained as they drove the moon car to the rim of North Ray Crater, 3.3 miles from the lunar vehicle *Orion*.

Young and Duke operated a portable magnetometer at several locations and recorded the first evidence of weak but unmistakable magnetism on the moon. The discovery implied the moon once had a molten interior like Earth's, and that it once spun much faster on its axis than it does today.

The Service Scientific Instrument Module with eight orbital science experiments is clearly visible in this view of the *Apollo 15* Command and Service Modules in lunar orbit. In the background is the lunar nearside, looking southeast into the Sea of Fertility. At bottom right is the crater Taruntius.

They packed away 213 pounds of rocks and launched *Orion's* cabin section after a stay of 71 hours, rejoining T.K. Mattingly in *Casper* for the trip back to Earth.

Apollo 16 lunar pilot Charles Duke stands beside the lunar roving vehicle at the Descartes landing site near Stone Mountain.

Earth, as seen by *Apollo 17* astronauts on their way to the moon. The view extends from the Mediterranean Sea area to the Antarctica south polar ice cap. Almost all of Africa's coastline is visible. The Asia mainland is on the horizon toward the northeast.

It was time now for Apollo's last hurrah.

For the astronauts to reach a valley named Taurus-Littrow, the Saturn 5 had to be launched at night, and shortly after midnight on December 7, 1972, darkness suddenly became daylight as the monster rocket roared to life.

Half a million viewers were rewarded by the most dazzling liftoff in the history of the spaceport, where more than 2,500 rockets had been fired in 22 years.

Three days later, Gene Cernan, Jack Schmitt and Ron Evans shot into lunar orbit aboard the command ship *America*. Schmitt, the only geologist to make it to the moon in Apollo, was agog as he rattled off descriptions of mountains, valleys, craters, rays and faults with an expertise not available to those who preceded him.

Harrison Schmitt uses an adjustable sampling scoop to retrieve lunar samples at the Taurus-Littrow site.

After Cernan and Schmitt skillfully landed the lunar ship *Challenger* at Taurus-Littrow, they stepped outside, the 11th and 12th Americans to walk the moon. As an excited Schmitt gazed around, he commented, "It's a geologist's paradise if I ever saw one."

Over three days they made three excursions in their moon car.

"Man is this mountain big," Schmitt exclaimed as he sifted through the rubble of an ancient landslide that brought rocks down from several layers of the mountain, and thus from several layers of lunar history. Some of the rocks were as large as a two-story house.

The geologist-astronaut described rock formations which indicated he and Cernan had tapped a lode of very old material, perhaps dating back 4.6 billion years, a time when many scientists believe the moon was formed.

At the end of their last excursion, Schmitt was first up the LM ladder. As Cernan, the last man of Apollo to touch the moon, started up, he paused for a final look at the bleak beauty around him, and said, "As I take these last steps from the surface for some time into the future to come, I'd like to record that America's challenge of today forged man's destiny of tomorrow. And as we leave the moon and Taurus-Littrow, we leave as we came, and, God willing, we shall return, with peace and hope for all mankind."

On December 19, the last men of Apollo came home, ending a brilliant chapter in the history of adventure and exploration.

More than two decades later, scientists around the world still are examining, and will continue to examine for decades, the treasure of six moon landings as they seek to learn how the moon, our Earth and solar system evolved.

What have they found so far?

When Neil Armstrong and Buzz Aldrin landed on the Sea of Tranquility, they found a lifeless world, baked by day and frozen by night. They found no winds, no storms, no clouds, no water, and only the thinnest of atmospheres.

Armstrong, Aldrin and the 10 men who followed them to the moon in the next three and one-half years returned with 842 pounds of lunar samples. They left behind science recorders that sent data for many years on moonquakes, meteor hits, radiation and the solar wind.

So far, only 20 percent of the lunar material has been examined by scientists in laboratories around the world. Most of the rocks are preserved in a pristine state in secure, hurricane-proof vaults in Houston and in San Antonio. They are kept in a pure nitrogen atmosphere to avoid contamination, to be held for the day when more sophisticated analysis techniques might better unlock their secrets.

What has emerged so far is a picture of a moon that was born in searing heat, lived a brief life of boiling lava and shattering collisions, then died geologically in an early, primitive stage.

It came into being some 4.6 billion years ago when great masses of gaseous matter called the solar nebula began condensing to form the sun, Earth and other planets and moons of our solar system. The nebula first condensed into chunks of space debris — from small pebbles to miles-wide boulders — then crashed together and fused to form celestial bodies.

In the case of the moon, this compacting of debris generated intense heat, which turned the lunar surface into a sea of molten lava, to a depth of several miles. When the lava cooled, this became the moon's primitive crust. Debris left over from the creation of the solar system continued to bombard the moon, carving out giant craters and valleys and forming mountains by piling up large piles of rocks.

The Apollo astronauts brought back at least a few fragments dating back 4.6 billion years, and several in the 3.8 billion to 4.5 billion-year range. These are important because they might reveal much about the early history of our Earth — a record that has been erased by weather, erosion, tidal action and the atmosphere. The oldest Earth rock found dates back only 3.8 billion years.

Lunar samples in MSC's Non-sterile Nitrogen Processing Line.

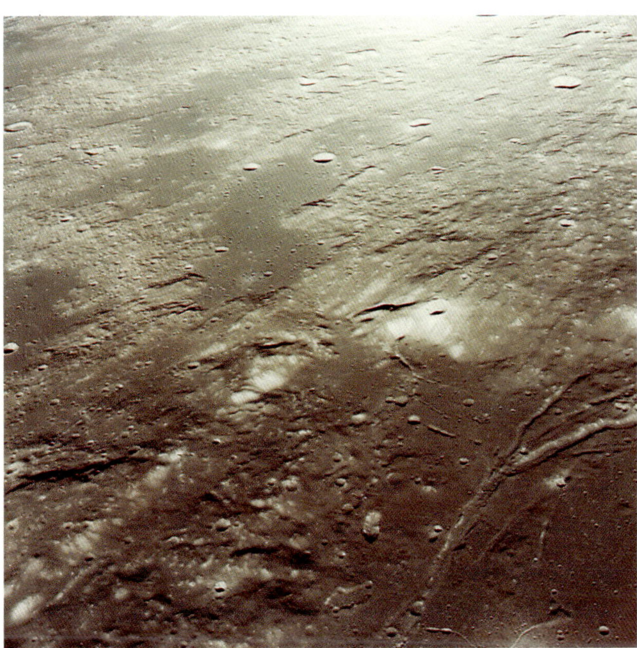

This *Apollo 17* moon view shows the slight orange cast identified by command module pilot Ronald Evans. Lunar module pilot Harrison Schmitt discovered an orange soil sample composed of fine glass particles rich in iron and titanium at Taurus-Littrow.

Most scientists believe the young Earth underwent the same period of meteorite bombardment and volcanism that the moon did for about half a billion years.

The weak gravity of the smaller moon could not prevent volcanic gases from escaping into space. But the Earth, with strong magnetic and gravity fields, held onto its volcanic vapors, and they formed an atmosphere and oceans, creating conditions for the development of life.

The moon became a dead body where life could not exist.

While research in the 1970s and 1980s concentrated on the chemical makeup and history of the moon, scientists now are also looking into how lunar resources could be used to help build a permanent moon base for humans.

One study is aimed at extracting oxygen from the lunar soil and converting it into breathing air for the base and for rocket fuel to carry explorers beyond the moon. Another looks at mining the moon for helium-3 or silicon, two substances that hold promise as energy sources. And there is a possibility of using the lunar mineral ilmenite to make concrete and other building materials for base structures.

As the quest for solutions to lunar mysteries continues, all the answers might not be known until others visit the moon. So far only six small sites have been sampled, leaving vast, unexplored regions like the lunar west and the hidden backside. They seemingly await the landing craft of another generation of explorers.

Chapter Twelve
'It's Been a Good Home'

Apollo was history. No longer was there the driving force that carried Americans to the moon. The United States had been technically challenged by a cold war rival, and a sense of fear and national pride had propelled it to respond, and to win.

The Soviets no longer were a threat in space, and the program slowed to a crawl. NASA's dreams of moon and planetary bases were doomed. Only the space shuttle remained. This huge, winged, reflyable spacecraft was a technological marvel — so much so that the program quickly escalated in cost and decelerated just as rapidly in its time schedule. Weeks became months, and developments expected to take months stretched into years.

It would be years before the shuttle would fly.

Something had to fill the gap. NASA wanted desperately to keep intact its management, astronaut, engineering and flight control teams.

The space agency had rockets and spacecraft left over from the three canceled Apollo moon missions. Engineers proposed modifying some of this hardware into a modest space station where astronauts could study the sun and other stars, conduct experiments seeking pure materials and medicines, and learn to live in space weightlessness for long periods in the event America one day decided to embark on a months-long manned expedition to Mars.

The Soviets had a space station, some argued. Could the U.S. do less?

The cost would not be great, and Congress and the administration agreed that NASA's teams were a great natural resource that should be preserved.

The project was named Skylab.

The third stage of a Saturn 5 rocket was stripped of its engines and converted into a complete station to be hurled into orbit by the first two stages of the big booster. *Skylab* was a "home away from home" with racks of scientific equipment, an astronomical laboratory, and more than 13,000 cubic feet for unparalleled comfort and freedom for three astronauts at a time. It

The *Skylab* space station in Earth orbit, with one solar array system wing fully deployed. A parasol solar shield shades the Orbital Workshop where the micrometeoroid shield and opposite solar array panel are missing.

had cooking facilities, a microwave oven, private quarters, a shower, exercise equipment and other luxuries unknown to the men of Mercury, Gemini and Apollo.

Skylab almost didn't make it. Barely 63 seconds after the Saturn 5 blasted off on May 14, 1973, ground controllers knew the project was in trouble. Radio signals indicated a combination meteoroid and heat shield had ripped away from the station, exposing it to the searing rays of the sun. The shield in flying off also tore away one of the laboratory's electricity-producing solar power panels and jammed the other against the side of the vehicle.

The first *Skylab* crew, Charles "Pete" Conrad, Paul Weitz and Dr. Joseph Kerwin, a physician, were to have blasted off the next day in a modified Apollo capsule to link up with and board the station. Their departure was delayed 10 days, while engineers sought ways to save the $2.6 billion project.

First priority went to cooling down the workshop, where temperatures inside soared to 125 degrees, too hot for astronauts to live in for very long. Experts devised a sun shade made of thin, aluminized Mylar and nylon. For freeing the stuck solar panel, various cutting and prying tools were made.

The first space salvage mission began May 25 with the liftoff of the three astronauts atop a Saturn 1B rocket. They hoped to repair the station and inhabit it for 28 days.

"We're on our way!" commander Conrad shouted as they shot into orbit and started a series of maneuvers to track down the 85-ton laboratory 275 miles above the Earth. Six hours later they moved in on their target and inspected the damage. They confirmed one solar panel was gone and the other was deployed a few inches, hung up by a one-inch wide, two-foot-long piece of aluminum, a remnant of the meteoroid shield.

While Mission Control evaluated the report, Conrad, Weitz and Kerwin docked to one end of the 118-foot-long station and ate dinner in their Apollo cabin. Controllers then gave them the go-ahead to try to free the panel.

Conrad undocked and maneuvered the Apollo ship to within 10 feet of the panel. Wearing a bulky space suit, Weitz opened the hatch and, with Kerwin hanging on to his legs, he leaned out and attempted to jerk the stuck panel loose with a long-handled tool resembling a boat hook. After an hour of pushing and tugging, Weitz conceded he couldn't break the aluminum strip.

The astronauts again docked with *Skylab* and slept overnight in their cabin. In the morning they removed connecting hatches and transferred for the first time into the sweltering laboratory. Their first business was to erect the umbrella-like sun shade. They pushed the folded device through a small scientific airlock module with a telescopic pole, then opened it — like an umbrella — into a 22-by-24-foot sheet that covered the area exposed by the ripped-away meteoroid shield.

Almost immediately, instruments in Mission Control and in the station noted a gradual drop in temperature in the workshop interior. Within days it dropped from 125 degrees into the 70s.

The astronauts then put the space station in order, unpacking supplies and setting up scientific instruments. Kerwin organized his medical laboratory.

Skylab 2, first of three manned missions, lifts off from Cape Kennedy May 25, 1973, with astronauts Charles Conrad Jr., Joseph P. Kerwin, and Paul J. Weitz.

They aimed a giant array of telescopes at the sun, pointed photosensors at the Earth, and conducted metal processing experiments in a small furnace. Kerwin supervised the drawing of blood, the monitoring of bodily functions and medical workouts on a stationary bicycle, treadmill and other devices.

But without power from the solar panels, *Skylab* ran on batteries only and the spacemen had to curtail many of the experiments. When two of the 18 batteries failed, the situation became more critical. Conrad suggested they think again about freeing the stuck solar panel.

"I think we have the tools to cut that strip or pry it loose," the commander said, and recommended a space walk. After evaluation and testing on the ground, Mission Control said okay.

Attached to 60-foot lifelines, Conrad and Kerwin work outside *Skylab*, using a cable cutter to cut the aluminum strapping holding the solar array system wing so that it would deploy.

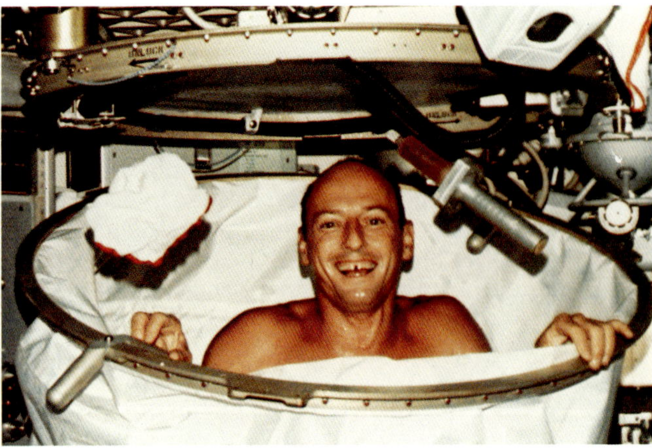

Conrad smiles for the camera after a hot bath in the *Skylab* shower. The shower curtain is pulled from floor to ceiling. Water comes through a push-button shower head attached to a flexible hose, and is drawn off by a vacuum system.

On the 13th day of the mission, Conrad and Kerwin, attached to 60-foot lifelines, stepped outside and assembled five sections of tubing into a 25-foot pole and attached it to a two-foot cutting tool resembling pruning shears. They made their way through a maze of trusses on *Skylab's* telescope mount to a point within pole's reach of the jammed solar wing.

It was difficult, but they finally broke the strip and then provided the muscle to erect the panel, both tugging on a rope they had attached.

Once the wing beam was out, three sheets of solar cells began extending from the edge. They caught the heat of the sun and converted it to energy to charge eight batteries which until then had been inoperative. The move added 3,000 volts of electricity to the laboratory.

"Wowee! Super-super," Conrad exclaimed.

"No more cold showers," Weitz added.

"Hot dogs instead of cold dogs," Kerwin quipped.

With the increase in electrical power, the astronauts settled into a daily routine of experiments. On the 21st day, using the solar telescope, Weitz snapped the first photograph ever of a flare eruption on the sun from above the obscuring veil of Earth's atmosphere.

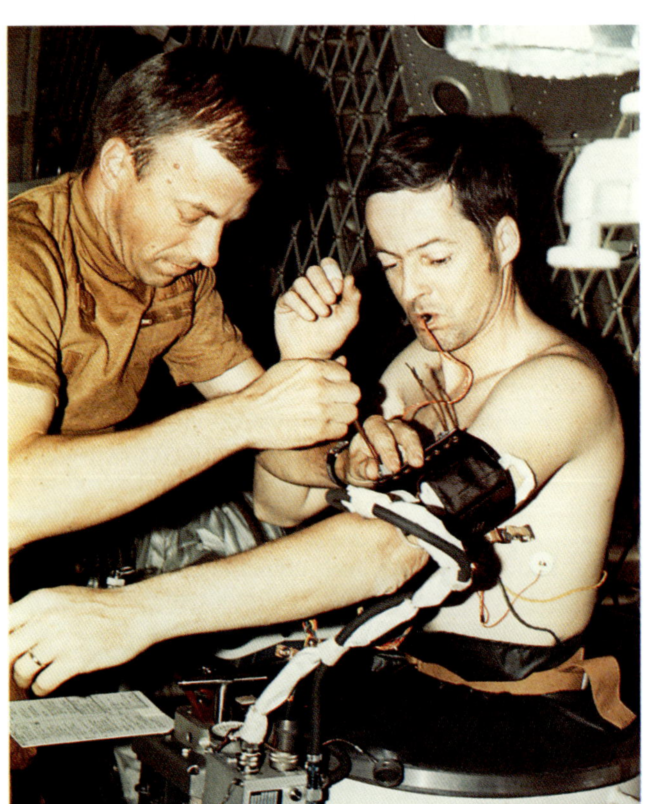

Kerwin serves as test subject for the Lower Body Negative Pressure Experiment. Weitz assists Kerwin with the blood pressure cuff.

The great solar eruption of June 10, 1973, seen in this spectroheliogram obtained during the first *Skylab* mission, extended more than one-third of a solar radius from the Sun's surface.

On the 28th day, they boarded the Apollo and headed back to Earth. On the deck of the recovery carrier all three showed some effects of the longest manned space flight yet. They walked unsteadily at first and doctors said they experienced dizziness, nausea and light-headedness. But within two days, all three were in excellent shape, having quickly readapted to Earth's gravity.

A month later, on July 27, a second astronaut team boarded *Skylab* and set up shop for a planned two-month mission. Five days later, serious trouble struck.

Alan Bean, Jack Lousma and Owen Garriott were suddenly awakened in the station by a master alarm in their attached Apollo ship. A fuel line had sprung a leak, and pressures and temperatures were falling rapidly in Quad Delta, one of four sets of Apollo steering rockets.

This was serious, because Quad Bravo had developed a leak earlier in the flight and had been shut off. That left only two quads to guide the craft through critical reentry maneuvers at journey's end. One was enough to do the job, but the flying task would be tricky.

Mission Control was concerned. Seventeen Apollo craft had flown in space, and there never had been a quad failure. What was causing the leaks? Was there a common fault that could spread to the other two quads and to the large main engine needed to brake Apollo out of orbit? Fuel for all these engines came from the same source. Was it contaminated?

Controllers considered a quick return to Earth. But that idea was dropped because *Skylab* itself was in good condition and could support the astronauts for several weeks until a rescue mission could be launched, if necessary.

NASA ordered emergency preparation of a rescue rocket at the Cape, with hundreds of engineers, technicians and managers working 12-hour shifts around the clock to get it ready for a launch as early as September 5. Astronauts Vance Brand and Don Lind trained to fly the rescue craft, an Apollo modified to carry five men.

Analysis determined there was no contamination in the batch of fuel from which the crippled Apollo drew its supply. Study of radio data showed there was no connection between the two quad failures.

Chris Kraft, director of the Johnson Space Center in Houston, radioed the astronauts, "We feel fairly confident that we've got two good quads for control. We're proceeding here as if we're going to have a normal mission." As a precaution, he said, rescue preparations would continue.

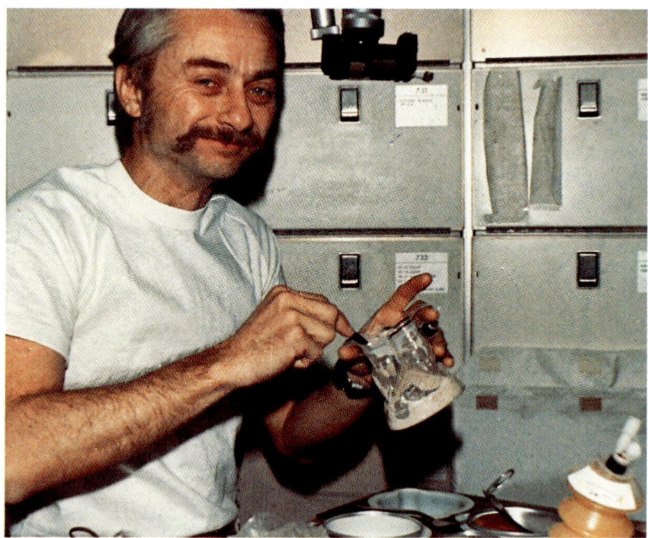

Skylab 3 science pilot Owen K. Garriott reconstitutes a prepackaged container of food at the crew quarters ward room table of the Orbital Workshop.

The last *Skylab* crew, launched November 16, 1973 atop a Saturn 1B, included astronauts Gerald P. Carr, Edward G. Gibson, and William R. Pogue.

"You said just the right words," said a relieved commander Bean.

The astronauts plunged into their experiments. Lousma and Garriott took a space walk on the 10th day to replace film in the solar telescope and to erect a second sun shade over the area on the station where the protective shield had peeled away. The shade erected by the first crew had begun to deteriorate because of constant bombardment from the sun's rays.

Two more space walks were conducted, one by Garriott and Lousma and one by Bean and Garriott. Both times the telescope film was changed, and on the second, a new set of gyroscopes was installed to keep *Skylab* on an even keel.

Otherwise the astronauts busied themselves with experiments, and each exercised at least an hour a day to keep heart and other muscles from deconditioning too much because of lack of use in zero gravity.

As the mission neared an end, controllers sent procedures for flying Apollo with only two good steering quads. They worked perfectly and the spacemen came safely home after a record 59 days in orbit. They all were a little dizzy at first, but recovered within a day.

The third and final *Skylab* crew boarded the station November 16, 1973.

Commander Gerald Carr and William Pogue concentrated on Earth resources cameras and sensors, surveying a wide area of the globe for hidden mineral or oil sources and other potential benefits. Edward Gibson, a solar physicist, pointed the six camera telescopes at the sun, stars and comet Kohoutek, a visitor from deep space whose sweep around the sun at that time was a scientific bonus for the crew. Medical experiments consumed several hours daily.

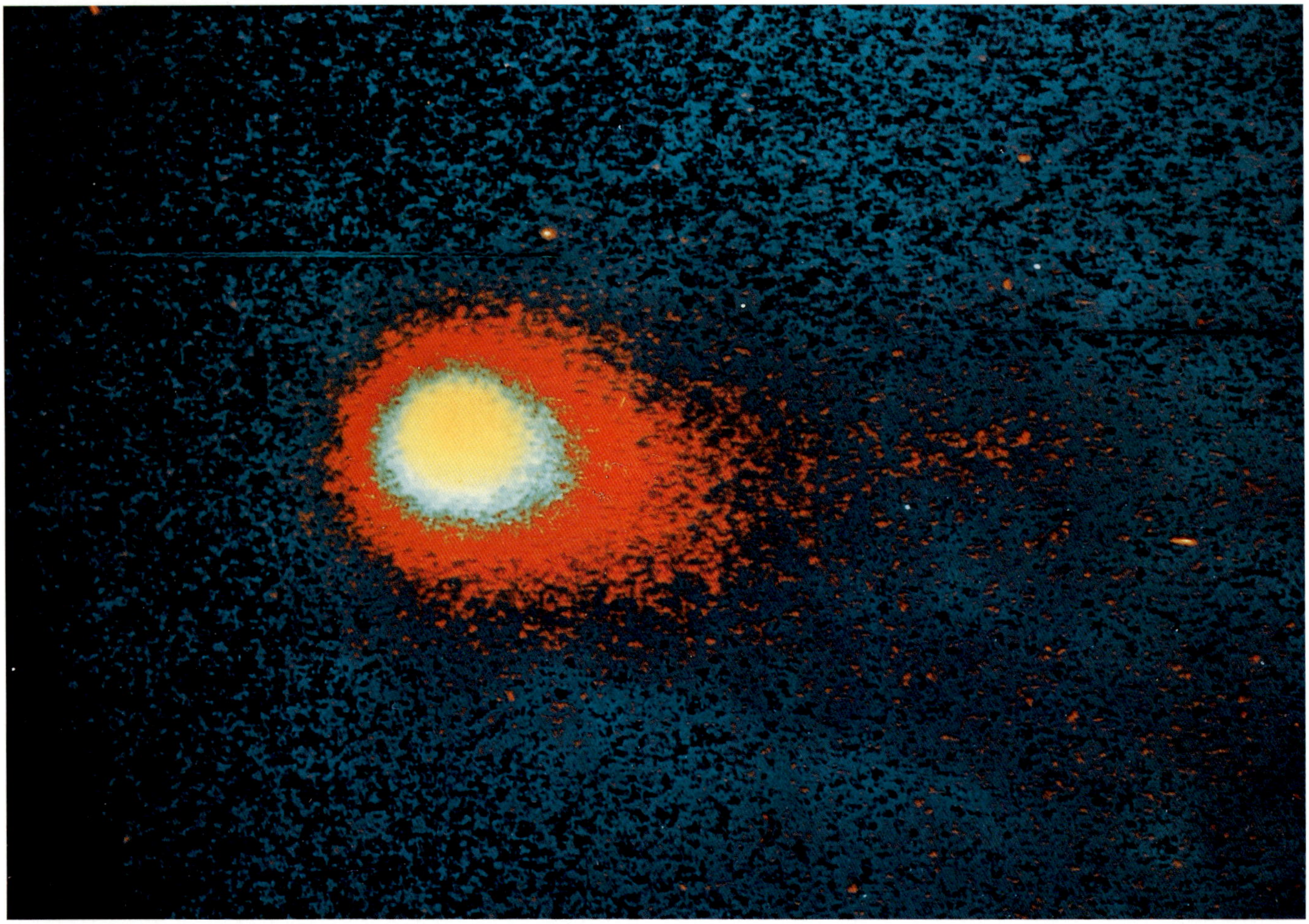

Comet Kohoutek, photographed outside *Skylab*, was described by Gibson as "one of the most beautiful creations I've seen."

The *Skylab* space station cluster in Earth orbit was photographed by the last *Skylab* crew during a final "fly-around" by the Command and Service Module before returning home. *Skylab* drifted through space for another five years before plunging to a fiery death through Earth's atmosphere.

The spacemen took four space walks, with the main purpose of each to change film in the telescope cameras. Twice two of the astronauts stepped outside to photograph Kohoutek. And although viewing the comet proved disappointing to earthlings, the astronauts had a ringside seat above the distorting atmosphere.

"I tell you it's one of the most beautiful creations I've seen. It's so graceful," Gibson remarked. Added Carr: "It's orange and yellow, just like a flame." They estimated the tail of gas and dust was more than eight million miles long, illuminated brightly by the sun.

After 84 days aloft, it was time to head home. As they separated the Apollo ferry ship, Gibson said, "It's been a good home. I hate to think we're the last guys to use it."

After splashdown, they stepped a bit shakily onto the carrier deck. But doctors said they were in excellent health, better than the two previous *Skylab* crews, thanks to a more vigorous exercise regimen.

The mountain of data returned by the three crews would tell humans much about themselves, their planet and their solar system.

Skylab drifted as a ghost ship through space for five years, with Earth's gravity gradually tugging it down until it made a fiery plunge through the atmosphere in 1979.

Chapter Thirteen

'Soyuz and Apollo Are Shaking Hands'

The five spacemen — three American, two Russian — each swallowed a pinch of salt, followed by a swig of vodka from a bottle passed around the table. They recorked the bottle, signed their names to it and vowed to drink the rest when they returned from the world's first international space flight.

The intake of salt and vodka was a Soviet custom signifying good fortune. Each chose his words carefully. No one drank a toast to the success of the joint U.S.-Soviet adventure. To the Russians, this was bad luck. "To toast our preparations, that is enough," explained cosmonaut Valeri Kubasov.

The five men were in the apartment of Alexei Leonov in Star City, home of the cosmonauts, near Moscow. Astronauts and cosmonauts were nearing the end of their final training for the flight. The table was heavy with food. Camaraderie abounded, just as it always had when these five had met over a three-year period to rehearse for the linkup in orbit of spaceships of the United States and Soviet Union — for 20 years keen rivals in the competition to conquer the space frontier.

Leonov would command the Soyuz. His equal on the U.S. side, commander of the Apollo ship, was

Crew members of the joint U.S.-U.S.S.R. Apollo-Soyuz Test Project docking in Earth orbit mission included (left to right) astronauts Donald K. Slayton, Thomas P. Stafford and Vance D. Brand, and cosmonauts Alexei A. Leonov and Valeri N. Kubasov.

Three Apollo-Soyuz Test Project engineers look over a Soyuz spacecraft docking system prior to an ASTP docking mechanism fitness test conducted at Johnson Space Center. Left to right are Robert White, American chairman of ASTP Working Group No. 3; Vladimir Syromyatnikov, his Soviet counterpart, and Yevgeniy Bobrov.

Cosmonaut Alexei A. Leonov, left, and astronaut Thomas Stafford train for the Apollo-Soyuz mission during a simulation exercise inside the Docking Module trainer at Johnson Space Center.

Tom Stafford, veteran of three earlier space missions. With Stafford were Vance Brand and Deke Slayton, the only one of the original Mercury astronauts who had never made a space trip and the only one of the seven remaining in the astronaut corps.

Slayton had been grounded for years by an irregular heart beat, which he managed to overcome in time to gain a seat on the Apollo-Soyuz mission. At 51, he was the oldest person yet named to a space assignment.

The toast in Leonov's apartment took place in April 1975. Launches of the two spacecraft was just three months away.

The seed for the joint flight was planted in 1969 by Thomas Paine, then the NASA administrator. Aboard the presidential jet Air Force One, flying across the Pacific to greet the returning *Apollo 11* astronauts, Paine suggested to President Nixon and his advisors that it might be desirable to develop international cooperation in space projects, especially with the Soviets. Nixon thought it was a laudable goal, that it could further ease Cold War tensions, and told Paine to begin negotiations with the Soviets.

In the aftermath of *Apollo 11*, in which the Americans demonstrated clear superiority in space, the Soviets seemed more willing to cooperate. Some critics suggested their motives were suspect, that by joining in a cooperative effort they would learn much about U.S. technology.

Whatever the motive, both nations established working groups, which quickly bought an idea put forth by Paine: Build a common docking device for the manned spaceships of each country as a means of facilitating a space rescue mission in case spacemen became stranded.

The docking device idea evolved into a plan to demonstrate that it would work by linking Apollo and Soyuz spacecraft in orbit.

The discussions culminated in the signing of a space agreement on May 24, 1972, at the Moscow summit meeting of President Nixon and Soviet leader Leonid Brezhnev. The linkup was scheduled for 1975.

For the next three years the American and Russian engineering teams modified their spacecraft, built and tested new equipment and learned each other's language. Astronauts and cosmonauts practiced orbital maneuvers and tests they would conduct in space and they each took more than 700 hours of language training. The rule in space would be: Cosmonauts speak English, astronauts speak Russian.

The history-making journey began July 15, 1975, with the launch of Leonov and Kubasov aboard the Soyuz from Baikonur Cosmodrome in central Russia.

At Cape Canaveral, it was 8:20 a.m. EDT, and Stafford, Slayton and Brand were still asleep. Their Apollo ship was scheduled for liftoff at 3:50 p.m. They were awakened at 9:10 a.m. with the good news

The American half of the Apollo-Soyuz Test Project, NASA's Apollo/Saturn 1B space vehicle, is launched from Kennedy Space Center at 3:50 p.m. EDT July 15, 1975.

An artist's concept of the Apollo-Soyuz docking.

that Leonov and Kubasov were in a solid orbit. During breakfast they watched a video replay of the Soyuz launch and then donned their space suits and headed for the launch pad.

Blastoff of the Saturn 1B rocket was on time and perfect, and the crew began a series of maneuvers designed to track down and catch Soyuz. Two days later they zeroed in on their target, and Stafford moved Apollo in for the linkup.

"I can see your beacon," Leonov radioed in English.

"Less than five meters distance. Three meters. One meter," the American commander called out. There was a slight shudder as the two ships, each traveling 17,400 mph, came together and the two halves of the new docking system engaged.

"Contact, capture," Stafford shouted. Switching to Russian, he said, "My spravilis (We have succeeded)."

"Well done, Tom," Leonov said. "It was a good show. Soyuz and Apollo are shaking hands now."

Three hours later, Stafford and Slayton were in a connecting tunnel, part of the docking mechanism, which also served as an airlock between spaceships with different cabin atmospheres. They closed the hatch at the Apollo end, and the Soyuz hatch was opened, exposing the beaming faces of the waiting cosmonauts.

"Tovarich (friend)," Stafford cried out as he shook hands with Leonov, who replied, "Very, very happy to see you."

Leonov gave both Americans the traditional Russian bear hug as they floated into the Soyuz cabin to be greeted by Kubasov. The spacemen traded gifts, including flags and commemorative plaques. Leonov, a gifted painter, gave the astronauts sketches he had made of them during training.

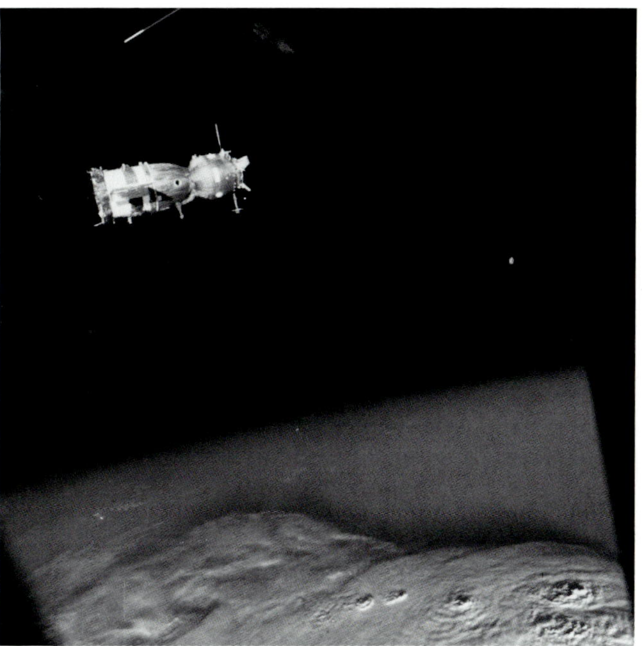

Apollo closes in on the Soyuz spacecraft as the two vehicles prepare to dock in Earth orbit.

They shared a meal, which included tubes of borscht, and then listened to radio greetings from their national leaders, Leonid Brezhnev and Gerald Ford.

Later, all five men gathered for a meal that included grilled steak in the Apollo cabin. Space fliers never have considered their food to be the best, prompting Leonov to remark, "As the philosophers say, the best part of a good lunch is not what you eat, but with whom you eat."

From top, astronaut Donald K. Slayton, cosmonaut Alexei A. Leonov and astronaut Thomas P. Stafford are photographed in the Soviet Soyuz Orbital Module while docked with Apollo in Earth orbit.

During the 47 hours the ships were linked, astronauts and cosmonauts moved easily between the two craft, to conduct joint scientific and medical experiments and to give earthlings televised tours of the United States and Soviet Union below.

In a news conference from space, answering questions from reporters at the two control centers, Slayton said that "through space flight men of various nations can gain a greater sense of understanding and cooperation." Leonov termed the flight "a first step on the endless road of space exploration."

On July 19, Apollo separated from Soyuz. A short time later the two ships redocked, and undocked again, with Soyuz the active partner this time, Leonov in control. This completed the total testing of the linkup device.

"Mission accomplished," Leonov declared as the two ships went their separate ways. "Good show," replied Stafford.

Soyuz returned to Earth two days later. Apollo stayed up three extra days for experiments and splashed down in the Pacific on July 24.

The landing appeared normal to television viewers. But inside the cabin, the three astronauts were involved in a tense, potentially deadly drama that neither Mission Control nor the world would know about until later. As Apollo descended, nitrogen tetroxide gas from steering jets filtered into the cabin, causing the astronauts to cough and their eyes to smart.

The crew, for an unknown reason, had forgotten to flip two switches that would have cut off the steering jet system and which would have activated the automatic landing system, including deployment of the parachutes. The gas was being drawn in through an air intake which opened at 24,000 feet.

The astronauts discovered the mistake when small stabilizing parachutes failed to open. They quickly unfurled them manually and did the same with the three main chutes. Stafford quickly shut off the steering jets, but by then a lot of yellow-brown gas was in the cabin.

Because of the late parachute opening, Apollo hit the water harder than usual and flipped over, leaving the three men hanging upside down in their couch harnesses. Brand hung limp and unconscious, a victim of the gas. Stafford scrambled to retrieve gas masks which he and Slayton donned. They put one on Brand and he recovered in minutes

The craft's ventilation system quickly cleared out the gas, and the astronauts righted the spacecraft by inflating flotation bags.

By the time a helicopter ferried them to the recovery carrier, all three were feeling good and they did not think about reporting the incident until hours later. Mission Control was not aware of what happened because of poor communications during the descent.

Stafford, Slayton and Brand told about the gas during a medical examination aboard the carrier, and doctors ordered them to sick bay for examination. They said the effects of breathing nitrogen tetroxide do not manifest themselves for a day or two.

The medics said there was a danger of lung damage and possible death if the three had inhaled a great amount of gas. The next day they complained of chest discomfort and they were confined to an Army hospital in Honolulu for observation and drug treatment. They improved, their spirits were good, and after three days they were released.

In the months that followed, the astronauts and cosmonauts toured both countries and were hailed by enthusiastic crowds. Visiting Star City, they drank to future joint missions, finishing off the bottle of vodka they had recorked three months before liftoff.

Despite all the glowing words before, during and after the mission about how Apollo-Soyuz would open the door to extensive space cooperation between the two superpowers, that promise soon faded as détente began to cool in 1976.

Cooperation lapsed for nearly two decades. But today, the breakup of the Soviet Union and the end of the Cold War have made it possible for the United States and Russia to cooperate with other nations to build an international space station, soon to orbit the Earth.

Almost overlooked in the excitement of Apollo-Soyuz was the fact that this was the last flight of an Apollo spacecraft, symbol of a triumphal era in which Americans landed on the moon, visited a space station and took part in a joint mission with old rival Russia.

It also marked the last flight of an American spacecraft built to land on water. Future ships would touch down on land.

It was time to move on to that next generation vehicle — the space shuttle.

Chapter Fourteen

'A Fabulous Flying Machine'

Years were required to solve the problems that beset the space shuttle, as engineers pushed the technological envelope to develop main engines capable of being used over and over, the solid fuel rockets to provide the initial push toward orbit, the computers and the fragile thermal tiles to protect the space plane as it plunged through reentry heat toward a runway landing.

One by one the problems were overcome as NASA made plans to build five of these revolutionary new space vehicles, each intended for 100 or more flights. They were named for historic sailing ships of exploration: *Columbia, Challenger, Discovery, Atlantis* and *Endeavour.*

First on the firing line was *Columbia,* on April 12, 1981. It had been six years since Americans had been in space, and more than half a million people crowded the Kennedy Space Center area to watch the sleek ship blast off from a refurbished launch pad that had once been the departure point for Apollo moon missions.

Commanding the shuttle on its maiden trip was veteran astronaut John Young, making his fifth space voyage. With him was rookie Robert Crippen, who would fly aboard the shuttle four times before retiring as an astronaut.

Space shuttle *Enterprise,* riding piggyback atop a modified Boeing 747, was used in NASA's approach and landing tests and a battery of pre-operational structural and dynamics evaluations. Astronauts Fred Haise (left, inset) and Gordon Fullerton, who conducted the first manned test of the orbiter atop the 747, use a "cherry picker" to leave the *Enterprise.*

In orbit, they checked *Columbia's* thousands of systems and opened the clamshell-like doors covering the shuttle's 60-foot-long cargo bay, capable of holding up to 65,000 pounds of cargo. Operation of the doors was critical. The inside surfaces function as radiators, ridding the spacecraft of excess heat. They must open before satellites can be launched from the bay or astronauts take space walks. And the doors must be closed

Space shuttle *Columbia* awaits its initial launch on KSC's Launch Pad 39A as two Gulfstream II planes, used as shuttle training aircraft, fly over the liftoff complex.

Wings of fire and feathers provide dramatic contrast as space shuttle *Columbia* and one of the graceful inhabitants of the ecologically rich Kennedy Space Center area both lift free of Earth's restraints, April 12, 1981. The maiden flight of the space shuttle, with astronauts John Young and Robert Crippen aboard, lasted 54 hours, ending at Edwards Air Force Base, California.

Astronauts John W. Young (right) and Robert L. Crippen train inside *Columbia's* cabin.

Earth serves as a backdrop for the shuttle's silhouetted vertical stabilizer and two orbital maneuvering system pods.

and locked firmly during reentry. Otherwise, the shuttle could go out of control.

Young and Crippen tested each door separately, opening and closing them. Then both doors together. "You're missing one fantastic sight," Crippen reported as the doors swung open revealing a deep black sky.

Because it was a first flight, NASA kept it short, just 54 hours, and to provide margin for error, scheduled the landing on the wide open dry lake bed at Edwards Air Force Base in California's Mojave Desert. No landings would be made on the main shuttle runway at Cape Canaveral, a narrow strip surrounded by alligator and snake-infested waters near the launch pad, until any possible bugs were resolved.

Commander Young fired braking rockets high over the Indian Ocean. Less than an hour later, *Columbia* glided unerringly toward the high desert, dropped its landing gear and touched down at 215 miles an hour, slightly faster than a commercial jetliner.

"Welcome home, *Columbia!* Beautiful! Beautiful!" Mission Control radioed.

"It is a fabulous flying machine," Young reported.

All previous manned spacecraft, U.S. and Soviet, had parachuted back to Earth, never to be flown again.

Columbia's cargo bay doors were opened and closed, and a solar radiator was deployed on the first day in orbit.

Columbia touches down on the dry lake bed at Edwards Air Force Base, California.

On *Columbia's* second trip to Earth orbit, astronauts Joe Engle and Richard Truly tested the Canadian-built remote manipulator system (RMS).

President Ronald Reagan and the First Lady greet astronauts Thomas Mattingly right, and Henry Hartsfield after the successful landing of *Columbia*, July 4, 1982, at Edwards Air Force Base following its fourth space venture.

Within days, *Columbia* was carried back to the Kennedy Space Center, bolted to the back of a modified 747 jetliner, to be groomed for another flight.

Officials scheduled three more test flights for *Columbia* before assigning it operational missions.

On November 12, 1981, it became the first spaceship to make a return trip to orbit, with a pair of rookie astronauts, Joe Engle and Richard Truly, aboard.

Trouble struck just two hours after liftoff. One of the craft's three power-producing fuel cells failed, and *Columbia's* electrical capacity was reduced by one-third. Mission rules dictated that if a fuel cell were lost, the crew would fly a so-called minimum mission of 54 hours instead of the planned five days. Loss of a second cell would mean an immediate return to Earth.

Engle and Truly worked tirelessly to conduct as much of their planned workload as they could. Their main task was to test the shuttle's $100 million, Canadian-built mechanical arm. On future flights, the 50-foot robot would be used to park satellites in orbit or to pluck them out of space if they required servicing or return to Earth for repair. They extended the six-jointed arm, bending and flexing it expertly.

The fuel cell problem was fixed and *Columbia* was back in space again four months later, on March 22, 1982, with astronauts Jack Lousma and Gordon Fullerton.

They further exercised the ship's systems and used the robot arm to lift an experiment package out of a berth in the cargo bay and place it back down again.

Lousma and Fullerton were to have landed at Edwards after seven days, but late winter rains in the desert forced them to touch down on Day 8 on another wide-open wasteland, New Mexico's White Sands Missile Range. NASA was not ready yet to commit to a Florida landing.

Columbia's final test flight began June 27, 1982, with astronauts T.K. Mattingly and Henry Hartsfield. They flew a near-flawless seven-day mission and on July 4, Independence Day, swooped to a landing at Edwards. Leading the cheers of half a million flag-waving greeters in the desert was President Ronald Reagan.

NASA administrator James Beggs declared the shuttle operational and said it was ready to carry cargo and passengers on a regular schedule into space.

To underscore that message, the agency dispatched its second shuttle, *Challenger,* to the Kennedy Space Center. Hitchhiking atop a Boeing 747, the new ship, built in California, passed low over the runway where *Columbia* had just landed and then above a reviewing stand where the president led the crowd in singing "God Bless America."

The era of manned commercial space transportation service began November 11, 1982, when *Columbia* sped into orbit for a fifth time, carrying in its cargo bay two communications satellites for paying customers, Satellite Business System's SBS-3 and Telesat Canada's Anik C-3.

Crew members were commander Vance Brand, pilot Robert Overmyer, and two members of a new breed of astronauts called mission specialists, Joe Allen and Bill Lenoir.

Nine hours into the flight, Allen and Lenoir, after a countdown, sent SBS-3 spinning out of the cargo bay. After Brand and Overmyer had guided *Columbia* to a safe distance of about 20 miles away, a rocket ignited and sent the satellite barreling to a stationary orbit 22,300 miles high.

To celebrate, the crew displayed a sign for television viewers: "Ace Moving Co. We Deliver."

The next day, the mission specialists successfully

Vance Brand, crew commander of *Columbia's* fifth Earth orbit, holds a sign that refers to the successful deployment of two commercial communications satellites. Other crew members, clockwise from Brand, include William Lenoir, Robert Overmyer, and Joseph Allen.

Sally Ride, first American woman in space, and fellow *Challenger* crew members Frederick Hauck (left) and Norman Thagard, prepare a meal. The tall experiment in the background is the continuous flow electrophoresis system.

dispatched Anik C-3 and Allen declared, "We're two for two."

Columbia had put the United States in the space transportation business, and it was time for a break. The ship went into drydock for several months for removal of test equipment, enlargement of its cabin, and upgrading of its engines.

Challenger stepped to the plate and made its debut April 4, 1983, roaring into orbit with the most sophisticated communications satellite ever built.

The $135 million Tracking and Data Relay Satellite (TDRS) was built to transmit as many as 300 million bits of information per second, equal to a 140-volume encyclopedia. NASA planned a series of three such satellites to relay communications and data between the Earth, its orbiting shuttles and as many as 25 other satellites, eliminating the necessity of a costly system of globe-circling tracking stations.

While Paul Weitz and Karol Bobko flew the shuttle, mission specialists Story Musgrave and Donald Peterson deployed the satellite. Troubles developed later as TDRS sped toward its outpost 22,200 miles high. But experts at NASA and TRW, Inc., the payload's builder, devised a thruster-firing scheme that guided it to a proper orbit.

The astronauts turned their attention to science experiments and one of the mission highlights, America's first space walk in more than nine years. Musgrave and Peterson floated into the cargo bay through an airlock, and for nearly four hours they tested new space suits and wrenches, winches and other tools designed for weightless mechanics. "Everything's working fine; it's really neat out here," Musgrave reported.

Challenger made its first return to Earth, in California, after a five-day journey.

Sally Ride made history during *Challenger's* next flight, which began June 18, 1983. After sending 82 men into space in 22 years, the United States finally put a woman in orbit on its seventh shuttle mission, as a member of the five-person crew. She was one of eight women on the astronaut team, now totaling 81. All of the women were selected by NASA in 1978 and 1979. Many more would be added in later years and routinely fly on the shuttle.

Ride and mission specialist John Fabian successfully released commercial communications satellites for Telesat Canada and the Indonesian government, and then used the robot arm to lift a West German experiment package out of the cargo bay and deposit it in its own orbit.

Story Musgrave, making the first space walk in nine years, moves down *Challenger's* payload bay door hinge line with a bag of latch tools.

A camera onboard the free-flying SPAS-01 satellite captured this picture of the Earth-orbiting *Challenger* against the blackness of space.

Commander Robert Crippen and pilot Rick Hauck used the free-flying package as a rendezvous target until Ride grabbed it with the arm and reberthed it in the cargo bay to be used on a later flight. The fifth crewman, Dr. Norman Thagard, performed medical experiments on himself and the others.

After six days in space, Ride remarked, "It's the most fun I ever had in my life."

Two months later, America's first black astronaut, Guion Bluford, Jr., flew aboard *Challenger* as part of a five-man crew that deployed a communications satellite and conducted experiments, including a study on how six rats adapted to weightlessness.

A refurbished *Columbia* was back in space in November 1983 with the first non-astronauts, including the first foreigner launched in a U.S. spaceship. They were Byron Lichtenberg, a biomedical engineer from Massachusetts Institute of Technology, and Ulf Merbold, a West German physicist.

They were the first of a new breed of space fliers called payload specialists, selected from outside the astronaut corps for specific missions — in this case, to work their specialties on Spacelab, a 23-foot, 17-ton cylindrical laboratory carried in the cargo bay. The 10-nation European Space Agency built the $1 billion workshop and donated it to NASA, with the provision that Europeans could work aboard it.

Guion Bluford, America's first black astronaut, checks one of the control knobs on the continuous flow electrophoresis system experiment.

The six crew members, the most yet for a shuttle, split into two teams and worked round-the-clock shifts in Spacelab. They conducted medical, astronomy, Earth survey, atmospheric and materials processing experiments during nine days aloft.

Flight 10, featuring *Challenger* and a crew of five was filled with high and low points.

The standout was a stiff, toy-like figure that appeared to hang suspended in space, silhouetted against a deep black sky, with a bright-blue, cloud-mottled Earth curving far below. This was a man, a human satellite, the first person to fly free in space, without a lifeline attached to his spaceship.

Astronaut Bruce McCandless was 320 feet from the shuttle, controlling himself perfectly by firing small jets that spit bursts of nitrogen gas from a pack on his back. "Beautiful...We sure have a nice flying machine here," he reported.

Using the nitrogen-propelled manned maneuvering unit (MMU), Bruce McCandless makes the first-ever space walk without restrictive tethers and umbilicals.

He thoroughly tested the jet-pack, called a Manned Maneuvering Unit (MMU), by turning somersaults while streaking 17,400 mph, 170 miles above his home planet. He moved back to *Challenger's* cargo bay and turned over the Buck Rogers-like pack to fellow spacewalker Robert Stewart, who maneuvered it through a series of tests and pronounced it "great."

While exercising the MMU they practiced techniques and tested tools that the next shuttle crew would need to repair a crippled satellite.

The successful space walks erased some of the disappointment the crew and NASA felt when two communications satellites deployed for Western Union and the Indonesian government failed to achieve their proper orbits because their onboard engines failed. It was not the astronauts' fault, but they, nevertheless, were not happy.

The two satellites would be retrieved and returned to Earth for repair on a later shuttle trip, and would be returned to space.

The flight ended on a high note when *Challenger* became the first shuttle to return from flight on the 15,000-foot concrete runway built near the launch site at the Kennedy Space Center. A landing at Kennedy saves NASA about $1 million and valuable time, because the winged spaceship does not have to be ferried from Edwards Air Force Base on a Boeing 747.

Flight directors Randy Stone, left, and Gary Coen, center foreground, watch the first space shuttle landing at Kennedy Space Center on a monitor in Mission Control at Johnson Space Center. Astronauts John Blaha and Guy Gardner man the spacecraft communicator console in the background.

The first shuttle repair mission took place in April 1984, after commander Bob Crippen and pilot Dick Scobee guided *Challenger* to within 200 feet of the Solar Max sun-study satellite, which had been disabled in space for more than three years.

While the chase was on, mission specialist Terry Hart, using the robot arm, dropped off an 11-ton research satellite, which was to be retrieved by another crew after a year.

Once *Challenger* and Solar Max were flying formation, mission specialists George Nelson and James van Hoften donned space suits and slipped through an airlock into the open cargo bay for the capture and repair assignment.

Nelson strapped on the MMU jet-pack and slipped over to his target, moving in to attach a docking device to a six-inch pin on the satellite. But the device failed to clamp, despite several attempts. With his jet-pack fuel running low, Nelson returned to the shuttle.

After thinking about it overnight, Mission Control told the astronauts to try to snare Solar Max with the robot arm. Crippen and Scobee steered to within 30 feet, and with one flex of the arm, Hart reached out, grabbed the satellite and brought it into the cargo bay.

Nelson and Hart were outside the next day, and during four hours of electronic surgery on Solar Max, they replaced a failed control system and a faulty electronics box. Engineers on the ground checked the revitalized satellite and pronounced it fit. Hart lifted it overboard with the arm and released it into orbit,

George Nelson and James van Hoften repair the faulty attitude control module on the Solar Maximum Mission satellite in *Challenger's* cargo bay.

where it began providing scientists with valuable information about the sun.

The first shuttle service call in space had been a tremendous success, but it had not been easy.

Discovery joined the fleet on August 30, 1984, on a satellite deployment mission with a crew that included the second American woman astronaut, Judy Resnik, who remotely unfurled a gold-colored solar sail so that it hovered 102 feet above the cargo bay in a test of a possible energy source for future space stations.

The *Discovery* crew tested the extended solar array as a method of using the sun to supply power to the space station.

The Satcom communications satellite spins from its protective shield in *Columbia's* cargo bay and begins its rise into space.

With three shuttles now operating, NASA began realizing a dream of routine access to space with reflyable spaceships. There were some problems along the way. Computers detected something amiss and halted a countdown at the last second, shutting down the main engines after they ignited; there were minor mechanical problems in orbit, and weather often halted launch attempts and diverted landings from Kennedy to Edwards.

While three fellow crew members sleep, five of *Challenger's* international crew pose in Spacelab during this November 1985 flight. Counter clockwise from Bonnie Dunbar are Steven R. Nagel, Reinhard Furrer (Germany), Henry W. Hartsfield, and Nubbo J. Ockels (The Netherlands). Not pictured are Ernst Messerschmid (Germany), James Buchli and Guion Bluford.

Astronaut Dale Gardner, wearing the MMU, captures the Westar IV satellite, which was returned to Earth in *Discovery*.

But the shuttles flew, and flew. Their crews deposited satellites, retrieved and repaired others, surveyed the Earth and stars, tested space station assembly techniques, conducted extensive medical research, and dropped off payloads and performed experiments for the U.S. military. Astronauts from Canada, France, Saudi Arabia, West Germany, the Netherlands and other countries flew, as did payload specialists from many disciplines.

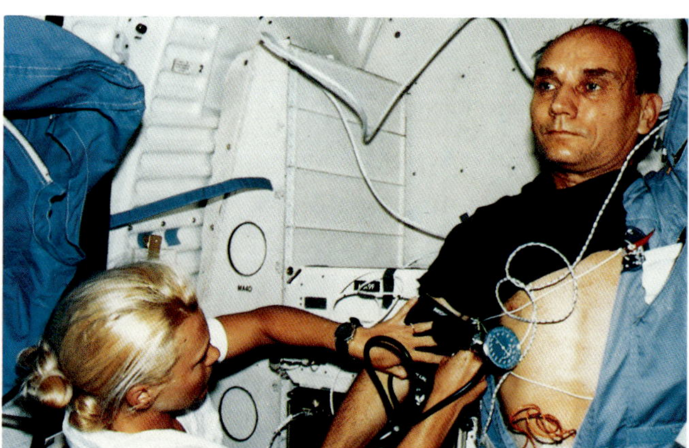

Dr. Rhea Seddon takes U.S. Senator Jake Garn's blood pressure during a medical experiment aboard *Discovery* in April 1985.

Even Congress got into the act when Senator Jake Garn of Utah, chairman of a subcommittee that oversaw NASA's budget, cajoled the agency into assigning him to a shuttle flight, the ultimate congressional junket. His counterpart in the House of Representatives, Bill Nelson of Florida, demanded equal respect, and he, too, made a trip into space.

Atlantis joined the fleet in 1985, a year in which NASA launched a record nine shuttles. There would have been 10 missions, but a December launching of *Columbia* was delayed several times by mechanical and weather problems and it became the first flight of 1986, in mid-January.

Anchored to foot restraints on RMS, astronaut Jerry Ross fits together aluminum tubing to form a beam that could serve the same structural function in a space station that a girder does in the construction of a steel-beamed building. Ross and astronaut Sherwood Spring easily fashioned a 45-foot beam out of 93 tubular struts that snapped together like Tinkertoys.

A nighttime lift-off puts space shuttle *Atlantis* in the right orbit to deploy communications satellites for Australia, Mexico and RCA in November 1985.

This all-military crew aboard *Discovery* delivered a secret military satellite for the Defense Department in January 1985. They are (front, left to right) Air Force Lt. Col. Loren Shriver, Navy Capt. Thomas Mattingly, (back, left to right) Air Force Maj. Gary Payton, Marine Lt. Col. James Buchli and Air Force Maj. Ellison Onizuka.

Flushed with success, and with a growing list of commercial and scientific customers seeking space on the shuttle, the space agency set an ambitious schedule of 15 launches for 1986 as it worked toward an ultimate goal of 24 a year.

Pressing the schedule, NASA set its second 1986 flight for January 24, just six days after *Columbia's* return.

Chapter Fifteen

'A Major Malfunction'

Challenger's crew of five men and two women gathered in astronaut crew quarters, seven miles from the shuttle launch pad, for an early pre-launch dinner. They had been through this ritual three times in the last four days, only to have their liftoff scrubbed by weather or technical troubles.

The crew for the 25th space shuttle mission was unique in that it featured America's first "citizen in space," Sharon Christa McAuliffe, a 37-year-old social sciences teacher from Concord, New Hampshire. She had been selected over 11,000 other applicants in the NASA Teacher in Space Project. Millions of school children waited eagerly to share in lessons she planned to teach from orbit.

Her crewmates were commander Francis R. "Dick" Scobee, 46; pilot Michael J. Smith, 40; Ellison S. Onizuka, 39; Ronald E. McNair, 36; Judith A. Resnik, 36, and Gregory R. Jarvis, 41, a Hughes Aircraft Co. satellite engineer. Scobee, Onizuka, McNair and Resnik had flown on earlier shuttle missions.

While McAuliffe was the attention-getter, the astronauts also planned to deploy two satellites, a Tracking and Data Relay payload and a package to study Halley's Comet.

Teacher-astronaut-in-training Sharon Christa McAuliffe (center) and her backup, Barbara Morgan, experience weightlessness aboard a KC-135 aircraft while monitoring an experiment involving magnetic effects — one of the tests that was to be performed on the *Challenger* flight.

Challenger's crew included (left to right, front) pilot Michael J. Smith, commander Francis R. "Dick" Scobee, and mission specialist Ronald E. McNair; (back) mission specialist Ellison S. Onizuka, payload specialist and "Teacher in Space" Sharon Christa McAuliffe, payload specialist and Hughes Aircraft Co. satellite engineer Gregory Jarvis, and mission specialist Judith A. Resnik.

They were set for another try the next morning, and their major dinner topic was a weather forecast that definitely was not favorable.

A rare winter freeze had invaded Florida, and meteorologists predicted a mass of cold arctic air would drop temperatures below freezing, to as low as 23 degrees, with a hard freeze extending from about 2 a.m. to 11 a.m.

Mission rules forbade a launch at temperatures below 31 degrees. That meant, if the forecast was accurate, it would be impossible to meet the ideal launch time of 9:39 a.m. But with an afternoon forecast in the 40s, NASA managers agreed the situation was marginally acceptable and ordered a go-ahead for another launch attempt the next day.

An extensive freeze protection plan was initiated to prevent water pipes on the launch pad from bursting in the cold. Steady streams of water flowed through the pipes, and thousands of gallons of antifreeze were poured into sound suppression water troughs beneath *Challenger's* engines.

But there was concern. A small group of engineers at Morton Thiokol, maker of the solid fuel booster rockets, alerted their managers that the cold weather could seriously affect the joints between rocket segments. They were worried about the synthetic rubber O-rings designed to seal the joints and prevent hot gases and flames from escaping. On several flights, primary O-rings had suffered severe hot gas erosion, and in a few cases minor erosion was found on secondary rings. The lower the outdoor temperature, the greater the erosion, with 53 degrees the lowest temperature to that time.

As early as the previous August, officials from Morton Thiokol and NASA's Marshall Space Flight Center briefed managers at NASA headquarters on the history and potential of the problem. They did not recommend halting flights, contending that continuing to fly was an acceptable risk while the joints were being redesigned.

At a series of teleconferences throughout the night, the outnumbered Morton Thiokol engineers argued their case, but Morton Thiokol and Marshall officials said their evidence was not conclusive and ordered the countdown to proceed. No top level NASA managers were informed of the concerns or the discussions.

Shortly after 2 a.m., the launch team was ready to begin pumping more than half a million gallons of liquid hydrogen and liquid oxygen into *Challenger's* massive fuel tanks. By now the temperature had dipped below freezing and an ice team was sent to assess conditions on launch pad 39B. The team found that the freeze protection plan had failed, that great quantities of water from burst pipes had leaked all over the pad, forming ice sheets and hundreds of large icicles.

Charles Stevenson, head of the ice team, told launch controllers, "Well, I'd say the only choice you got today is not to go. We're just taking a chance hitting the launch vehicle." A decision was made to start fueling and then send the ice team back out after sunrise.

Stevenson returned at 7:30 a.m. and found conditions to be even worse. His team used long-handled tools to break off some of the biggest icicles and to crack ice that formed in the sound suppression troughs despite the antifreeze.

A rare winter freeze had invaded Florida, and despite an extensive freeze protection plan, water pipes on the launch pad burst in the cold. Leaking water formed ice sheets and hundreds of large icicles.

There was concern that chunks of ice would break off during the force of liftoff and possibly hit and damage sections of the shuttle. NASA engineers ran computer projections, calculating the paths the ice would take, and concluded that with current wind conditions, none of the large chunks would strike the spaceship.

Shortly before 9 a.m. the astronauts, one by one, climbed into *Challenger's* cabin and were strapped in. Their families were escorted to a viewing area three miles away.

The countdown ticked down to nine minutes, where an extended hold was called to wait for a bright sun to warm the weather. Shortly after 11 a.m., the launch control center informed the astronauts the count would resume shortly. The temperature outside was 36 degrees. "All right!" commander Scobee exalted.

A final poll of all monitors found no problems and the signal was given to start the count. Liftoff was just nine minutes away, at 11:38 a.m.

"Here we go," pilot Smith shouted as *Challenger* thundered off its pad and headed for space. The initial burst from the engines shook loose chunks of ice, and some hit the left booster rocket, but did no serious damage.

But something else occurred when the solid fuel rockets ignited. Unseen by anyone except launch pad cameras, small puffs of black smoke issued from the lower joint on the right booster. As the joint rotated ever so slightly, opening a tiny gap, the primary O-ring was too cold to seat and seal immediately. The cold-stiffened putty in the joint collapsed and hot gases

Challenger's crew wave to onlookers as they leave the Operations and Checkout Building on their way to Pad 39B. Front to back, they are Dick Scobee, Judy Resnik, Ron McNair, Michael Smith, Christa McAuliffe, Ellison Onizuka, and Greg Jarvis.

Challenger lifted off from Pad 39B January 28, 1986, at 11:38 a.m. EST with a crew of seven astronauts and the Tracking and Data Relay Satellite. An accident 73 seconds after liftoff claimed both crew and vehicle.

with a temperature of nearly 6,000 degrees rushed past an area of the primary ring. Because of the joint rotation, the second O-ring was not in its sealed position and the gases burst through.

The puffs of smoke continued for two and one-half seconds, then stopped when aluminum oxides from the burning fuel miraculously plugged the leaks in the joint before the flame could escape.

Unaware of the danger, the astronauts were excited as *Challenger* climbed higher and faster. "Go, you mother," Smith cried out. Twenty-two seconds after liftoff, Scobee throttled down the main engines, holding down the energy while the shuttle pushed through an area called "Max Q" — maximum aerodynamic stress — where a combination of high winds and severe dynamic pressures buffeted the vehicle. After 36 seconds, Scobee restored full power, confirming, "Go at throttle up."

Shear winds up to 84 miles an hour, some of the highest ever for a shuttle flight, blasted *Challenger* as it dashed through Max Q, prompting Smith to remark, "Looks like we've got a lot of wind up here today." When the winds hit the right booster rocket, they jarred loose the aluminum oxides that had sealed the lower joint.

Nothing now could hold back the fire, and flames burst like a blow torch through the joint opening 58 seconds into the flight. They burned through and collapsed the lower half of the external fuel tank, where liquid hydrogen that fed the main engines was stored. When the lower strut that attached the booster to the tank broke away, the rocket swiveled on its upper strut and its nose crashed into the top of the tank, releasing liquid oxygen to mix with the hydrogen to create an awful fireball in the clear sky.

Smith, suddenly aware something terrible was happening, uttered the final words preserved on a spacecraft cabin recorder, "Uh Oh!" Some of the crew futilely activated emergency oxygen supplies. But *Challenger,* thrown free of the fuel tank and boosters by the force of the explosion, tumbled out of control and was ripped apart by tremendous aerodynamic forces that medical experts believed killed the seven crew members within seconds.

"Obviously a major malfunction," said a mission commentator reporting from Mission Control in Houston. "We have no downlink. We have a report from the flight dynamics officer that the vehicle has exploded."

While the families of the astronauts, tens of thousands of other spectators and television viewers watched in horror, the two booster rockets broke loose and flew crazily through the sky until destroyed by a radio signal from a range safety officer. Chunks of fiery debris fell into the Atlantic Ocean. The crew cabin remained relatively intact but shattered when it hit the water.

Challenger died 8.9 miles up after a flight of just 73 seconds.

The nation was in shock as, over and over again,

"The vehicle has exploded!"

people watched reruns of the tragic explosion on television. To Americans, space flight was matter of fact. It had been going on for 25 years; a generation had grown up knowing that humans went into space, walked on the moon and returned safely to Earth. Hardest hit were the millions of school children who watched in classrooms and auditoriums as the teacher who was to instruct them from orbit died before their eyes. Their grief became their parents' grief. Many required special counseling.

President Reagan postponed his State of the Union message, scheduled that night, and appeared on television to praise the astronauts. "The crew of the space shuttle *Challenger* honored us by the manner in which they lived their lives," he said. "We will never forget them, nor the last time we saw them, this morning as they prepared for their journey and waved goodbye, and 'slipped the surly bonds of Earth to touch the face of God.'"

Navy and Coast Guard vessels began the grim task of recovering the debris, most of it scattered over hundreds of square miles of the ocean bottom. The shattered crew cabin was not found until nearly six weeks after the accident.

Wreckage from space shuttle *Challenger* was displayed in the Logistics Facility at Kennedy Space Center to assist the process of finding the cause of the accident.

Reagan named a presidential commission to investigate. In June, after exhaustive hearings, it issued a report that cited troubling lapses in judgment, expertise, communications and management. It called the explosion "an accident rooted in history" and said NASA and its contractors had accepted the growing risks "because they got away with it the last time."

The commission concluded the direct cause was a leak in the right booster joint that allowed flame to escape and listed cold weather as a contributing factor. "The space shuttle's solid rocket booster problem began with the faulty design of its joint and increased as both NASA and contractor management first failed to recognize the problem, then failed to fix it and finally accepted it as an acceptable flight risk," it stated.

The commission also determined the space agency was under too great a pressure to launch and was stretching its capabilities in scheduling 15 shuttle flights in 1986. It made nine major recommendations, including redesign of the booster rocket joint and other shuttle system improvements, formation of a NASA safety office, overhaul of shuttle management, improvement of internal communications and reduction of the flight rate. The three remaining shuttles were grounded while moves were made to implement the recommendations.

Management was shaken up at both NASA and Morton Thiokol.

To correct the rocket joint problem, engineers added a third O-ring, a metal capture ring to prevent joint rotation at ignition, improved insulation, heaters to warm the joints in cold weather and dozens of other changes. They also improved the shuttle's main engines, brakes and landing gear. *Columbia, Discovery* and *Atlantis* underwent a $2.4 billion overhaul, with a total of 56 major design changes and 400 lesser ones.

Although NASA had originally approved five shuttles, the fifth was on hold because of a lack of funding. After considering it for six months, President Reagan agreed to build *Endeavour* at a cost of $2.8 billion to replace *Challenger*.

At the same time, the president ordered NASA out of the business of launching commercial satellites and encouraged private industry to build unmanned rockets to do the job. Rocket builders like McDonnell Douglas, General Dynamics and Martin Marietta soon began offering launch services. The Pentagon also removed most of its military payloads from the shuttle manifest, assigning them to expendable rockets.

That sharply reduced the shuttle's space transportation role, confining it mainly to scientific and satellite repair missions.

NASA had plenty of time to assess this new role as engineers proceeded carefully and deliberately with the shuttle redesign.

The redesigned O-ring joint, post-firing test booster DM-8 in Utah.

Chapter Sixteen
'Piece of Cake'

Thirty-two months passed before a shuttle was ready to fly again. But fly it did. On September 29, 1988, *Discovery* thundered off its launch pad and into orbit with a crew of five seasoned astronauts.

"We sure appreciate you all getting us up in orbit the way we should be," Rick Hauck, the shuttle commander told the launch team. With him were pilot Dick Covey and mission specialists George Nelson, Mike Lounge and Dave Hilmers.

At an awards ceremony in the White House Rose Garden, President Reagan announced, "America is back in space again."

Six hours into the flight, Lounge and Hilmers released from the cargo bay a $100 million Tracking and Data Relay Satellite, a replacement for another TDRS lost in the *Challenger* accident. The new satellite soared into a stationary orbit 22,300 miles above the Earth, where it joined a sister satellite, launched earlier, to serve as a space relay point for transmitting radio signals between ground stations and the shuttle and as many as 25 other orbiting payloads.

The Tracking and Data Relay Satellite soars toward its stationary orbit 22,300 miles above the Earth.

During the remainder of the four-day mission, the astronauts conducted a dozen science and technology experiments and tested the more than 200 changes made to the shuttle as a result of *Challenger*. There were minor problems with a cooling system and a radio antenna, but, otherwise, Hauck reported, the "machine was absolutely superb in its performance and cleanliness."

During a news conference from orbit, the astronauts poignantly remembered the five men and two women who died aboard *Challenger*. In an emotional message that they took turns reading, they eulogized their fallen comrades as "fellow sojourners" and friends and expressed "reverence for those whose sacrifice made our journey possible."

"At this moment, our place in the heavens makes us feel closer to them than ever before," Nelson said as

Discovery lifts off September 29, 1988, putting America back in space. The Emergency Rescue crew and tank in the foreground are approximately one mile from the launch site.

the crew relayed a television view of *Discovery's* tail silhouetted against a striking picture of the Earth below.

"Lest we ever forget...that to ascend to this seemingly tranquil sea will always be fraught with danger, let us remember the *Challenger* crew whose voyage was so tragically short," Covey added.

"Today, up here where the blue sky turns to black," said Hauck, "we say at long last to Dick, Mike, Judy, to Ron and El, and to Christa and Greg: Dear friends, your loss has meant that we could confidently begin anew. Dear friends, your spirit and your dreams are still alive in our hearts."

The next day, *Discovery* glided to a landing on a dry lakebed at Edwards Air Force Base in front of 400,000 flag-waving spectators assembled to cheer America's return to space flight. NASA said shuttles would land in the wide-open California desert for the

Vice President George Bush greets *Discovery's* crew as they leave the shuttle after landing at Edwards AFB.

foreseeable future until new brakes and landing systems could be checked out. Only then would the spaceship resume landing on the confined runway at the Kennedy Space Center launch site.

Shuttle chief Richard Truly called the mission "an absolutely stunning success," but cautioned, "Even when we've flown a few flights, we are not going to forget the *Challenger* accident. For the people who work in the program, that's going to be on their minds for a long time."

The rebound from the explosion continued two months later with *Atlantis* soaring into orbit on the third shuttle mission dedicated solely to the Defense Department.

Although the Pentagon had directed, in the wake of *Challenger,* that military payloads be removed from the shuttle manifest, it would be a while before unmanned rockets would be ready to transport them, and the Defense Department had a huge backlog of critical satellites waiting to go. So during the next four years a total of eight shuttle military missions were launched until the final one was flown aboard *Discovery* in December 1992.

Pentagon officials drew a secrecy curtain around most aspects of these flights, including the type and purpose of the payloads, but in most cases, word leaked to the news media on the goal of each mission. The cargoes included satellites for missile warning, reconnaissance and communications.

Science began to dominate the manifest in May 1989 on the fourth post-*Challenger* flight, during which the crew of *Atlantis* deployed the shuttle's first planetary probe.

After being released from the cargo bay, a motor aboard the *Magellan* spacecraft sent it streaking through the solar system toward Venus, 158 million miles away.

"*Magellan* is deployed," commander Dave Walker radioed as the craft sped away.

"*Magellan* is coming to life and is on its way to Venus," Mission Control replied.

In August 1990, the 7,500-pound probe completed its 15-month outward journey and settled into orbit about Venus. For the next four years, its powerful radar peered through thick Venusian clouds and mapped more than 90 percent of the surface, sending to Earth images that clearly showed features the size of a football field, about 300 feet, with 10 times the clarity of any previous photos obtained from the planet.

The images should over the years help scientists learn much about the history of Venus and how it evolved so differently from its sister planet, Earth. It is believed both planets were almost identical when the solar system formed some 4.6 billion years ago, but Venus today has an atmosphere made up mostly of carbon dioxide trapped beneath thick clouds of sulfuric acid. Temperatures on the red-hot surface reach 900 degrees Fahrenheit.

Magellan ended its journey in 1994 when ground controllers commanded it to dive to a fiery death, gathering data about the Venusian atmosphere as it plunged toward the surface.

A second shuttle planetary mission in five months began October 18, 1989, when astronauts aboard

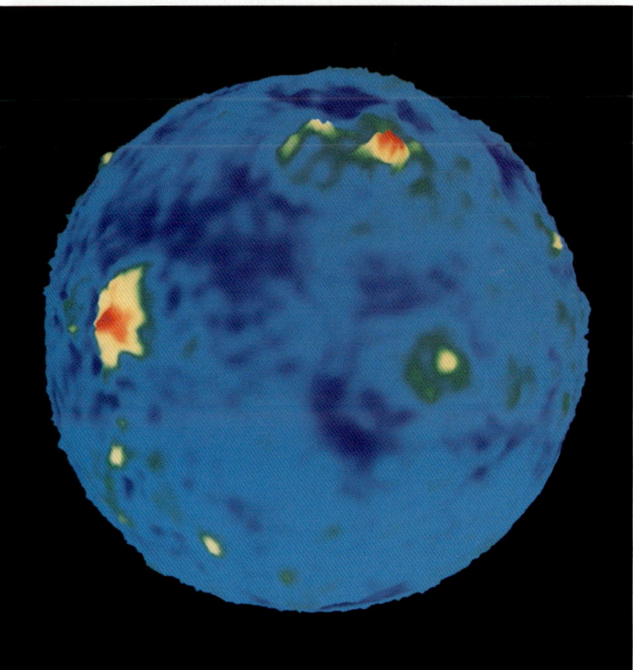

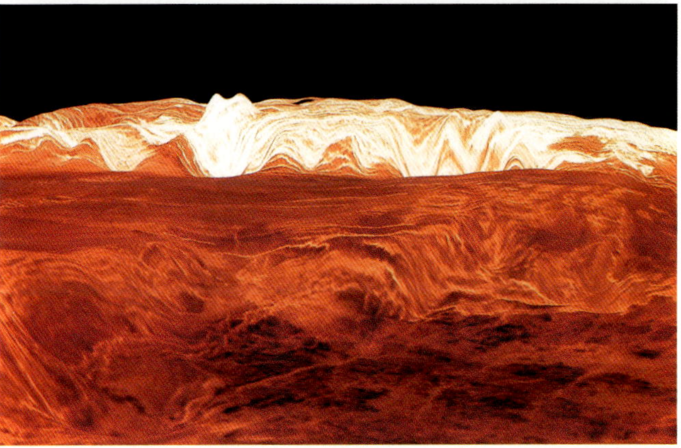

Top: An artist's rendering of *Magellan* at Venus. (JPL photo.) Center: A *Magellan* view of Venus. Surface topography is displayed as three-dimensional highs (red) and lows (blue). The highest region on the planet is Maxwell Montes, the yellow and red area near the top right of the globe. The high mountains of the Beta Regio (left center) are shown in red. *Bottom:* The simulated color of this *Magellan* view of Maxwell Montes on Venus is based on color images obtained by the Soviet Venera 14 and 15 landers.

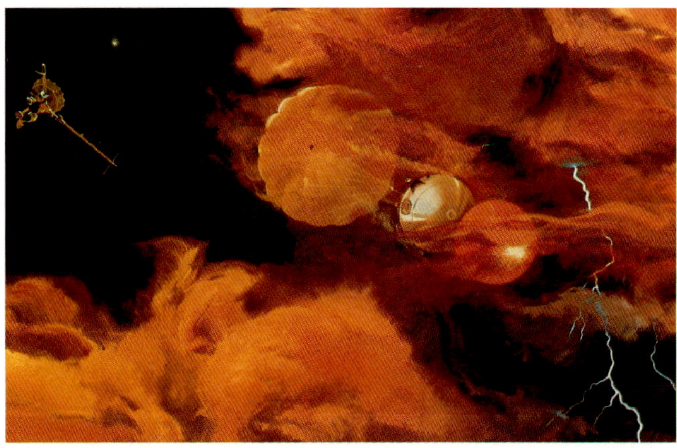

Galileo, deployed from *Atlantis* in October 1989, is expected to reach Jupiter in December 1995.

Ulysses, deployed from *Discovery* in October 1990, begins its five-year mission to the Sun.

The Gamma Ray Observatory, still on *Atlantis'* robot arm, is about to be deployed in April 1991. Astronauts Jerry Ross and Jay Apt manually extended GRO's high-gain antenna.

The solar array panel of the Upper Atmosphere Research Satellite is extended before the huge satellite is freed from the grasp of *Discovery's* robot arm and sent to a higher orbit.

Atlantis started the 6,700-pound *Galileo* spacecraft on a six-year, 2.4-billion-mile trip to Jupiter, the sun's largest planet.

When *Galileo* reaches Jupiter in December 1995, it is to fire a probe that will parachute into the Jovian atmosphere. Then the main body of the spacecraft will swing into orbit about the planet to make a detailed study expected to last about two years.

At a cost of $1.5 billion, *Galileo* was the most expensive unmanned space probe ever built, prompting a researcher to call it the "Rolls-Royce of spacecraft."

Because Jupiter has apparently undergone little change from its origin, the planet's composition is thought to resemble the cloud of gas and dust from which the sun and its planets were formed. "Being there and observing over a period of nearly two years will allow us to sort out the dynamics of Jupiter and its moons," said Edward C. Stone, a mission scientist.

Other major planetary craft launched from the shuttle in the following years were *Ulysses* (1990), which orbited the sun and explored its polar regions; the Gamma Ray Observatory (1991), to study space radiation, and the Upper Atmosphere Research Satellite (1991), which made the most extensive examination yet of Earth's troposphere, the upper level of the atmosphere.

The probe that received the most worldwide attention was the Hubble Space Telescope, set free in 1990 by a crew aboard *Discovery*.

Hailed as astronomy's most advanced observer, Hubble orbited 300 miles high, above the turbulent atmosphere, which blurs the vision of even the largest telescopes on Earth. Astronomers hoped to observe deep space objects 10 times more clearly than from the ground.

But blurred pictures sent by the telescope told mission scientists that mistakes had been made in manufacturing. Investigation disclosed that the telescope's main mirror had been ground to the wrong specifications, which meant that only a small fraction, about 15 percent, of the light gathered by Hubble was properly focused to a sharp point. The remaining 85 percent was spread out into a large, fuzzy halo and was essentially unusable.

Astronomers devised computer programs to remove this wasted light from the images, and the telescope produced spectacular images of dust rings in the centers of galaxies that may be hiding massive black holes from view, peered into the heart of clusters of stars and tracked a storm on Jupiter.

But because of the flawed mirror, many observations, especially those of faint, distant objects, could not be made.

In designing Hubble, NASA had recognized that all telescopes need periodic upgrades and had scheduled space shuttle astronauts to service the satellite every three years during its planned 15-year lifetime.

Above: During the third of five spacewalks to repair the Hubble Space Telescope, astronaut Jeffrey Hoffman signals directions to ESA astronaut Claude Nicollier (top right) who controls the robot arm from *Endeavour's* aft flight deck station. Hoffman and Story Musgrave earlier changed out the Wide Field Planetary Camera. *Bottom right:* Astronaut Kathryn Thornton hovers over equipment which she and Thomas Akers used to change out Hubble's solar array panels during the second space walk.

The first of these servicing missions had been planned for 1993, with the astronauts scheduled to replace a planetary camera, designed to look at relatively bright objects. After the mirror problem was discovered, designers realized they could polish the mirrors in the new camera to refocus the light properly.

But that would correct the vision problem for only one of Hubble's many instruments, so the servicing mission was delayed a few months while other solutions were sought.

Finally, on December 2, 1993, the shuttle *Endeavour* — the *Challenger* replacement, which had made its launch debut in 1992 — rocketed into orbit on the most difficult and challenging satellite repair mission ever attempted.

The veteran crew was commanded by Dick Covey and included pilot Ken Bowersox, payload commander Story Musgrave and mission specialists Jeffrey Hoffman, Kathy Thornton, Tom Akers and Claude Nicollier.

The flight captured the public's imagination like no space mission since the days of Apollo moon missions. Millions around the world watched extensive television coverage as four of the astronauts, working in teams (Musgrave-Hoffman and Thornton-Akers) took a record five space walks totaling 35 hours, 28 minutes.

The performances were flawless as Covey and Bowersox guided *Endeavour* to within 35 feet of Hubble, and Nicollier reached out with the shuttle's robot arm to snare it.

From flight days four through eight, the spacewalking teams, hanging on the end of the 50-foot arm and dangling from handrails on the four-story-tall, 13-ton telescope, completed one fix after another.

They installed a package as big as a phone booth containing a set of mirrors that served as eyeglasses to correct the nearsightedness; replaced the planetary camera; outfitted the observatory with a new set of power-producing solar panels; added three new gyroscopes, and replaced several other components needed to keep Hubble working.

"Piece of cake!" exclaimed Thornton, perched on the end of the robot arm, as she cast off one of the old solar panels.

Indeed, the work went so well that the space walkers finished most repair tasks ahead of time.

But it was several weeks before anyone knew if their job paid off, after scientists had time to fine tune and test the new instruments.

The answer came on January 13, 1994, when NASA and Hubble scientists faced a standing-room-only crowd of journalists at the Goddard Space Flight Center in Maryland. Two pictures of a star were distributed — one taken by Hubble before the repair, the other after it.

"The trouble with Hubble is over," said one of the participants.

The first photo showed a large, fuzzy halo, the second, a sharply focused star.

The scientists also revealed that for the first time, astronomers were able to follow the spiral arms into the innermost regions of the M100 galaxy. And

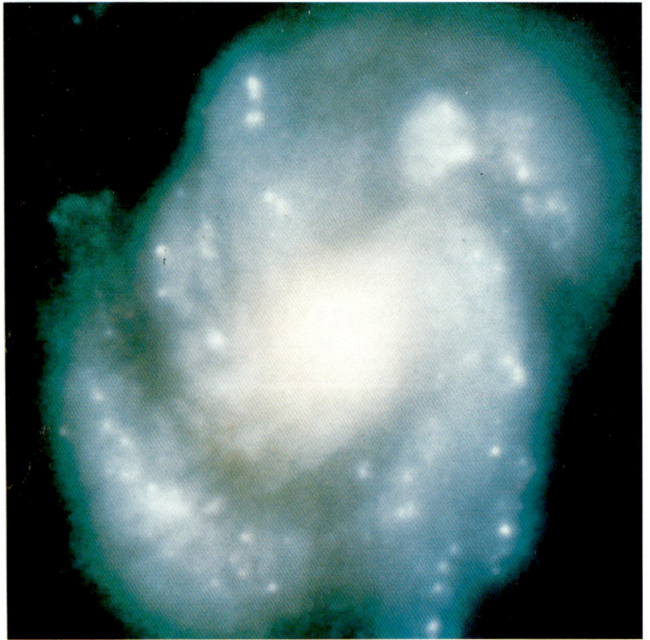

Hubble Space Telescope pictures of the M100 galaxy, before (left) and after HST repair.

because of the clearer images, they were able to see extremely faint stars in M100's outer regions that will allow them to accurately determine its distance.

Another Hubble image revealed thousands of stars in a cluster of young, hot stars, where before astronomers had only been able to detect the hundred or so brightest stars.

Hubble managers said the orbiting telescope was now every bit as good, even better, than it was intended to be when it was launched. For years, thousands of scientists will use the refurbished observatory to learn more about the universe. They will look for evidence of massive black holes in the centers of galaxies, continue the search for planets around other stars and begin to calibrate distances to faraway galaxies.

In the post-*Challenger* era, shuttle astronauts several times carried in their cargo bays laboratories called Spacelab and Spacehab, where they conducted experiments in metallurgy, astronomy, Earth observations and life sciences.

Many of these experiments were provided by other countries, and astronauts from several nations flew shuttle missions to help operate them. Among them were Japanese, Canadian, West German, Swiss, French and Italian astronauts.

Probably the most symbolic visitor was Russian cosmonaut Sergei Krikalev, who flew aboard *Discovery* in February 1994. The first Russian to be launched in a U.S. spacecraft presaged a new era of space cooperation in which Americans, Russians, Europeans, Japanese and Canadians would unite to build the world's first international space station.

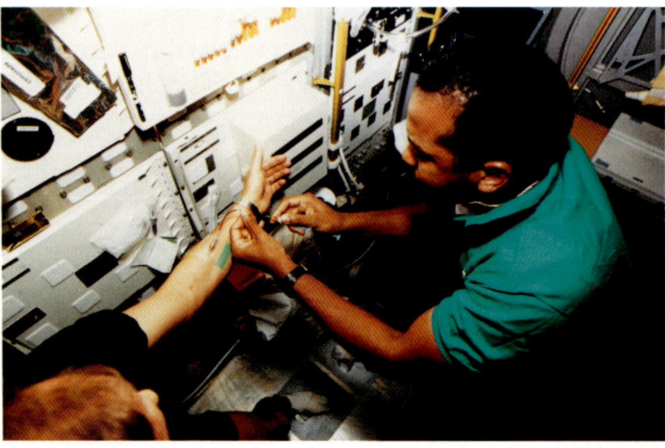

Hans Schlegel (left) serves as a test subject at the Anthrorack in the science module onboard *Columbia* during the April-May, 1993 mission. Astronaut Bernard Harris, a physician, performs one of many blood draws designed to help investigate human physiology under microgravity conditions. Schlegel was one of two payload specialists representing the German Aerospace Establishment (DLR) on the 10-day Spacelab D-2 mission.

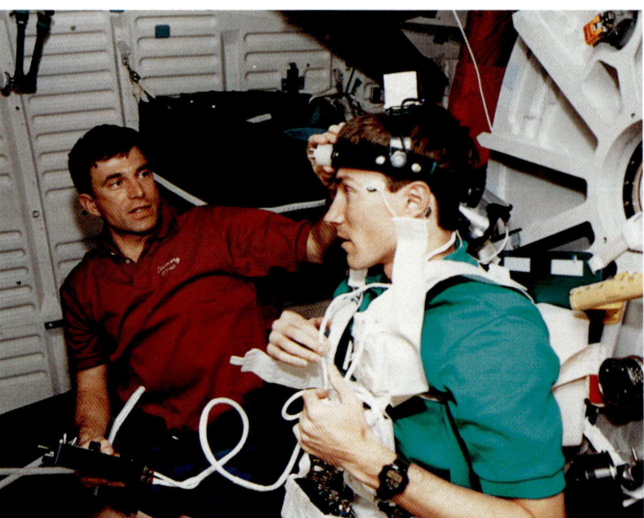

Astronaut Ronald M. Sega (left) and Russian cosmonaut Sergei K. Krikalev work on a joint U.S./Russian metabolic experiment on *Discovery's* middeck during the February 3-11, 1994 mission.

Chapter Seventeen

'A Force for Peace and Progress'

From its beginning, NASA dreamed of establishing a permanent space station in orbit where researchers, working in a gravity-free environment, could produce exotic materials and pure pharmaceuticals, where astronomers could gaze outward to study the universe from above the Earth's distorting atmosphere, where scientists could examine their own planet in depth and where humans could learn to live and work for long periods in space. But in the 1960s the focus was on the Apollo moon program, and in the 1970s and early 1980s, it was on developing and flying the space shuttle.

The space agency partially realized its dream in 1973, when it used hardware left over from canceled Apollo missions to orbit a modest, temporary outpost called *Skylab*. Over a nine-month span, three teams of three astronauts manned the station for periods up to 84 days, proving the importance of having people conduct science in space.

The Soviet Union, after losing the moon race to the Americans, concentrated all its effort in the 1970s on a space station, and in 1974 launched the first of a series of *Salyut* laboratories which were inhabited by rotating crews of two or three persons who orbited for months at a time.

NASA continued to push strongly for a permanent station of its own, and President Reagan in his 1984 State of the Union message ringingly endorsed the development and deployment of a station within a decade. The station, he said, would not only be a symbol of America's continued leadership in space but could also open a vast new commercial enterprise out there. "We can follow our dreams to distant stars," he declared.

But Reagan and his administration never followed up, did not provide the commitment necessary to carry out this grand plan. NASA, too, was remiss in failing to properly and vigorously sell this project to the public or to Congress.

Then came the terrible explosion of space shuttle *Challenger* and the loss of its seven astronauts on January 28, 1986. The space agency entered a long period of remorse, restructuring and rebuilding, and the space station project drifted aimlessly.

In 1986, after five *Salyut* stations, the Soviets launched their new, larger, much improved laboratory named *Mir 1* and cosmonauts inhabited it for up to a year at a time, gaining valuable experience and data for the time when Russians might embark on months-long trips to Mars or elsewhere in the solar system.

The Russian *Mir* station, photographed by *Discovery's* crew during the February 1995 rendezvous.

A space shuttle orbiter prepares to dock to space station *Freedom* in this painting done for NASA by artist Alan Chinchar. *Freedom* was to be operational by the mid 1990s.

Goddard Space Flight Center was given the responsibility to develop a space robot to aid in space station construction. This concept was developed by Martin Marietta.

The Space Station Emergency Rescue Vehicle, one of the directives resulting from the Presidential National Space Policy, would provide safe and reliable access to and from space station. This art concept is by David Russell, Omniplan Graphics.

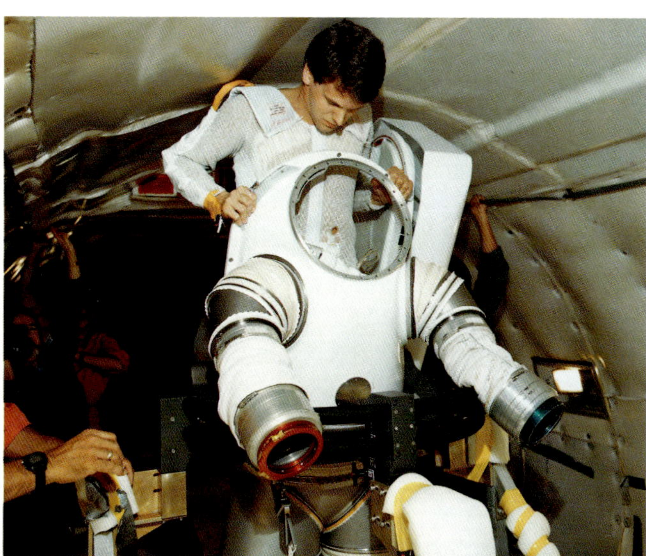

An early version of the ZPS Mark 3 space station spacesuit undergoes testing aboard a NASA KC-135 "zero-gravity" aircraft. Pressurized to 8 pounds per square inch, it would allow spacewalks on short notice without lengthy pre-breathe to purge nitrogen from crew bloodstreams.

Spurred by *Mir 1*, the United States signed an agreement with Japan, Canada and the member nations of the European Space Agency to jointly develop an international space station. The U.S. would maintain the leadership role and provide the major elements of the station, with the Europeans and Japanese building research modules and Canada developing a mobile service center, a maintenance depot and a large robotic arm.

They agreed to call the station *Freedom*.

While the foreign partners pushed forward with their parts of the agreement, NASA had continuing problems with Congress, with many of the members miffed at the escalating costs of the project. The original $8 billion price tag climbed to $14.6 billion, and that figure did not include the cost of the estimated 20 shuttle launches it would require to assemble the outpost. The National Research Council scrutinized the project and in 1987 estimated the eventual cost would reach $32 billion.

In the budgetary process, the station, designed for a 30-year lifetime, underwent approval for funding one year at a time. With each annual debate came new congressional calls to cut back costs, forcing NASA to redesign and downsize the station several times.

George Bush was elected president in 1988 and said he soundly supported the space station, a permanent moon base and a manned mission to Mars by the year 2019. But this vision, too, languished in an unenthusiastic Congress.

In 1992, Bill Clinton defeated Bush and directed a complete review of the space station program, asking NASA to present alternate designs.

By now the Soviet Union had broken apart and ceased to exist as a united country, and space officials of the cash-strapped Russian republic were vigorously marketing to the world the still-to-be-built *Mir 2* station. The most interested party was the United States.

While these two old rivals began negotiations on *Mir 2*, the space station survived in 1993 by only one vote in the U.S. House of Representatives. The project was saved only because of strong lobbying by a team led by NASA administrator Daniel Goldin.

After months of talks, the U.S. and Russian space agencies signed an agreement in June 1994 making Russia the newest partner on the international space station. The partners dropped the name *Freedom* and the project was referred to unofficially as International Space Station *Alpha*.

With Russia adding its *Mir 2* module, Soyuz transport capsules and launch vehicles to the program, the station grew larger by 25 percent and the number of crew it could accommodate increased from four to six.

Supporters of Russian participation cited the 22-year history of Russian space stations and the expertise that would provide. And they argued that making the Russians a peaceful partner was both sound foreign policy and sound economic policy.

The addition of the Russians reduced overall U.S. costs, and in the summer of 1994 the station received strong backing from both houses of Congress. NASA now expects steady funding of about $2 billion a year.

While American costs will fall in the long run, the U.S. agreed to pay the Russians $400 million from 1994 through 1997 in support of at least 10 space shuttle missions to *Mir 1* from 1995 to 1997. During these flights, shuttle crews will carry up new solar panels to replace aging arrays. Russia will add modules to its station to hold scientific equipment from both the U.S. and Russia. At least four American astronauts will conduct research aboard *Mir 1* for a total stay time of at least 24 months.

In the first of a series of cooperative missions, Sergei Krikalev flew on a space shuttle mission in 1994, the first of several cosmonauts scheduled to orbit in the American spaceship.

During the mission, President Clinton spoke with Krikalev and the five American members of *Discovery's* crew, telling them: "When we get the space station finished with the contributions of Russia, Canada, Japan, Europe and the United States, it's going to be a force for peace and progress that will be truly historic. And you will have played a major role in that."

Top: President Bill Clinton, U.S. Rep. Jack Brooks (D-Texas), left, and NASA Administrator Daniel Goldin, center, speak to orbiting *Discovery's* crew in February 1994, from a console in Mission Control. Aboard was Sergei Krikalev (bottom picture), the first Russian cosmonaut to orbit in the shuttle.

Crew members for the joint space shuttle/*Mir* missions meet the press at Johnson Space Center. Left to right on the dais are cosmonauts Nikolai M. Budarin, Anatoly Y. Solovyov, Gennady M. Strekalov and Vladimir N. Dezhurov, and astronauts Bonnie J. Dunbar and Norman E. Thagard.

Cosmonauts on *Mir* took this picture of *Discovery* as the shuttle approached within 37 feet of the Russian space station.

A year later, on February 5, 1995, another cosmonaut, Vladimir Titov, was aboard the shuttle *Discovery* when it flew to within 37 feet of Russia's *Mir* station and cruised alongside and around it 250 miles above Earth. It was the first convergence of manned American and Russian spacecraft since the Apollo-Soyuz linkup 20 years earlier and was a rehearsal for later shuttle-*Mir* dockings.

"Houston, this is the most beautiful thing I've ever seen in space," *Discovery* commander James Wetherbee radioed Mission Control as he eyed the huge Russian outpost. "Unbelievable."

Mir commander Alexander Viktorenko excitedly remarked, "It's almost like a fairy tale, almost too good to be true."

"The next time we approach, we will shake your hand and together we will lead our world into the next millennium," Wetherbee told his fellow commander.

Discovery's crew included Eileen Collins, the first woman to pilot an American spaceship, and Bernard Harris, the first African-American to walk in space.

The Boeing Co. was named lead contractor for the

A technical rendition of space shuttle *Atlantis* docked to the Kristall module of the Russian *Mir* space station. The joint U.S./Russian mission is scheduled for June 1995.

U.S., and around the world manufacturing of station segments began. Hundreds of engineers, technicians, managers, cosmonauts and astronauts in all the countries involved started learning the others' languages.

The shuttle trips to *Mir 1* mark the first of three phases planned in the project. In phase two, as now envisioned, Russia will launch a core *Mir 2* module in May 1997. That will be followed in quick succession by nine additional launches of Russian and U.S. modules and hardware, to be assembled by cosmonauts and astronauts. This initial station should be ready for a small crew of two or three researchers by December 1997. Early the next year, a Soyuz craft will be attached to the station and left there for use as an escape vehicle in case of trouble.

By the end of phase two, command of the shuttle and *Mir 2* will be handled by a unified control center consisting of both the U.S. and Russian mission control centers in Houston and Kalingrad. The Houston center will have overall command and control authority.

Phase two of the international space station, with elements provided by the United States and Russia comprising the human tended space station.

Phase three will include the addition of the Japanese research module in 1999 and of the European laboratory in 2000, along with other supporting hardware. The station should be completely assembled and ready to support a six-person crew in October 2001.

The researchers will have many comforts in their orbiting home, including a refrigerator, microwave, shower, stereo and exercise equipment, a small library and sleeping areas where they can escape for privacy.

These amenities should help as individuals spend months at a time searching the skies for secrets of the universe or scanning Earth for minerals, oil, fertile lands and pollution sources.

NASA is most excited about the unique environment the station will provide for research on the growth of protein crystals. Such information could greatly enhance drug design and research into the treatment of diseases. Crystals grown on much shorter space shuttle flights for research into cancer, diabetes, emphysema, parasitic infections and immune system disorders have been far superior to anything grown on Earth, officials reported.

International space station *Alpha*, in its completed/fully operational state, with elements from the United States, Europe, Canada, Japan and Russia.

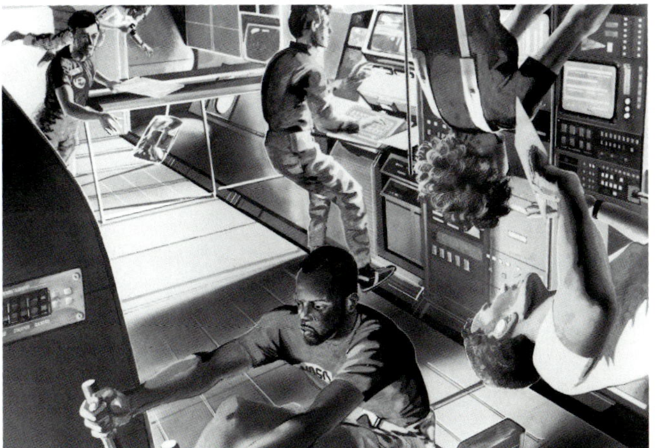

Space art by Harold Smelcer depicts typical scenes of an international space station crew living and working aboard *Alpha*.

The space station also will be a testbed for future technologies and a laboratory for research on new, high-technology industrial materials.

Daniel Goldin, the NASA administrator, remarked: "As the world redefines itself in the wake of the cold war, the space station is a catalyst for international co-operation and a powerful symbol of U.S. leadership in a changing world....The space station will inspire a new generation of Americans to explore and achieve, while pioneering new methods of education to teach and motivate the next generation of scientists, engineers, entrepreneurs and explorers."

This *Galileo* view of the moon, looking down on the moon's north pole, was assembled from 18 images taken as the spacecraft flew by the moon on December 7, 1992. Slicon and helium-3, two elements that could be mined from lunar soil, have potential as new energy sources for earthlings.

Chapter Eighteen

'The Economic Boon Will Be Enormous'

Tucked away in NASA closets and filing cabinets are yellowing blueprints for bases on the moon and trips to Mars.

These ventures have long been on the space agency's wish list, but lack of funds and a national commitment have kept plans on the shelf.

The current cooperation of many nations in building an International Space Station could open the way for the United States to join future multi-national efforts to build ever larger stations, to return humans to the moon and transport them to the Red Planet.

NASA's plans contain drawings for moon bases ranging from simple campsites for basic rock-gathering explorations to large lunar colonies for mining operations, astronomical observations and solar power plants.

If an international expedition to the moon is not mounted, Japan and the European Space Agency may strike out on their own.

In June 1994, ESA's science director, Roger Bonnet, unveiled an open-ended program to colonize the moon. It would begin early in the 21st century with exploration by robot orbiters and landers, followed by robots which would build a base, which Bonnet said could be ready for human occupation by 2020. The plan is under study but has received no major funding.

The Japanese see lunar exploration as an economic opportunity, and hundreds of engineers from the country's top construction companies are considering plans for elaborate mining operations on the moon by 2015. Conference centers, hotels, research labs and a lunar city would follow by 2050. The construction companies have formed a consortium called Construction Engineers Group for Aerospace Study (CEGAS).

These colonies would mine the lunar soil for two elements that potentially could provide earthlings with new energy sources and reduce or even eliminate their dependence on coal, oil and natural gas.

One of the elements is silicon. It could be used to make solar cells that astronauts could assemble into large solar panels. These solar panels would draw power from the sun and convert it to microwaves,

NASA's proposed Lunar Discovery Orbiter, depicted in this painting by Pat Rawlings of SAIC, is skimming just 20 kilometers over the moon's surface as it approaches Amundsen, a large terraced crater near the moon's south pole. LDO is expected to accomplish the first high-resolution global survey of the moon.

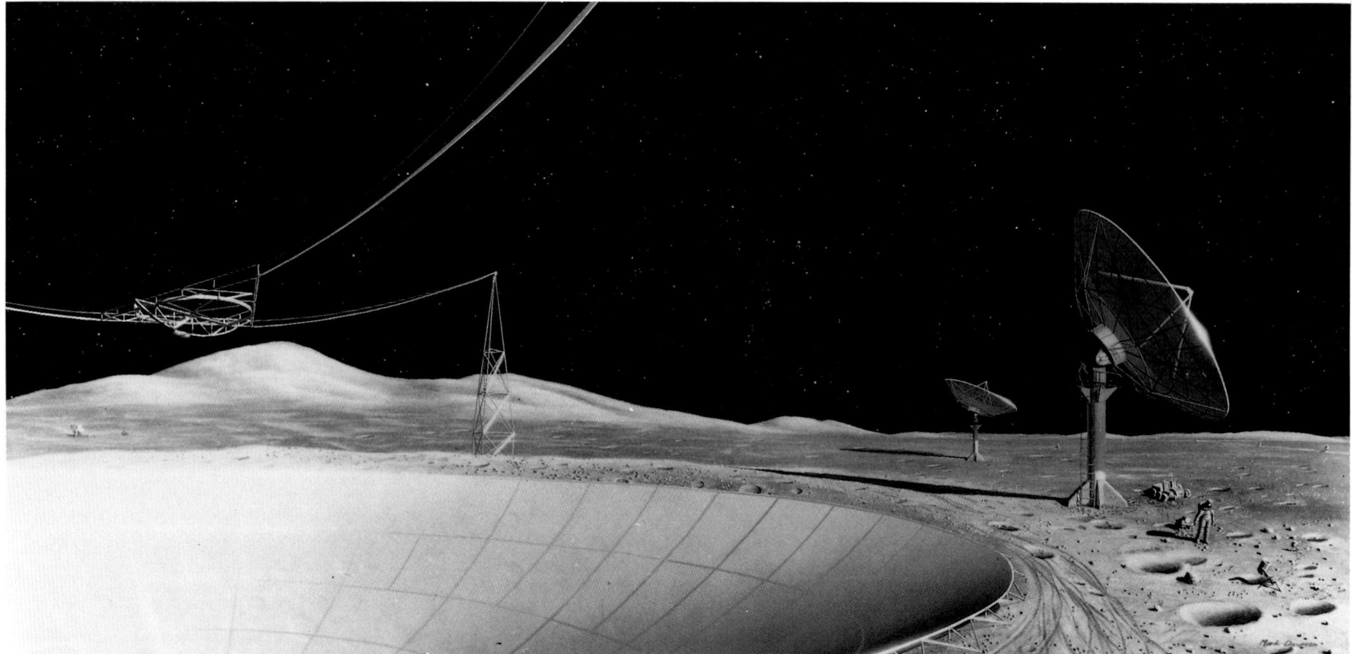

In this concept of a lunar observatory on the far side of the moon, a large fixed radio telescope is mounted in an existing crater. It focuses signals into the centrally located collector suspended above the crater. Two steerable radio telescopes are on the right.

which could be beamed to stations on Earth and converted to electrical energy. Such a system could provide inexhaustible, reliable and inexpensive power to a planet whose population is projected to double to 10 billion people by the middle of the 21st century.

The Japanese also are designing moon-roving robots to dig up tons of helium-3, a rare form of helium that Apollo astronauts found in abundance on the moon. Some researchers report helium-3, in short supply on Earth, could be used in nuclear power plants to safely produce electrical energy with little or no radioactive waste.

The economic payoff could be tremendous, says William Braselton, vice president for business development for Harris Corporation's Government Aerospace Systems Division, who long has advocated that the U.S. mine helium-3 on the moon.

"One space shuttle load of 25 metric tons of helium-3 would electrically power the United States for one year and have a market value of $75 billion today," Braselton said. He said the moon contains an unlimited supply of helium-3, which has been deposited on the moon's surface by the solar wind over the last 4.5 billion years.

NASA's shelved plans also call for mining the lunar soil for ilmenite, a source of lunar oxygen to be used to produce breathing air for a moon base and fuel for rockets to return to Earth or to set out for Mars or other destinations. Ilmenite also could be used in making concrete for lunar construction.

The moon's natural resources also could be employed in the manufacture of plastics, adhesives, industrial chemicals and other products.

"Lunar industrialization will equal new industries, new resources, new sources of energy and new jobs for our people," said Haym Benaroya, an associate professor of mechanical and aerospace engineering at Rutgers University in New Jersey.

"The economic boon will be enormous," he said.

NASA says a lunar outpost also could be the site of a powerful astronomy observatory. Astronomers long have been frustrated because Earth telescopes must deal with many obstacles that distort their views of the universe. Among them are the refractive interference of the atmosphere, wind-induced vibration and the bending of the telescope by gravity. A moon-based telescope would have a clear look at black holes, quasars, pulsars and other mysteries that reside out there.

If moon colonies are built, can tourists be far behind?

"How would you like to be one of the first holidaymakers on the moon?" asked Thomas Cook, the United Kingdom's leading travel agent in a press release in 1994. His company has been assembling a list of potential lunar travelers since man first walked on the moon a quarter century ago.

With nearly 2,000 persons booked, Cook in 1994 issued a new set of travel certificates and luggage labels to reconfirm these would-be travelers on the company's "Out of This World" holidays. Some applicants, eager to be among the first, have offered a cash deposit and credit card numbers.

The Japanese construction consortium CEGAS foresees a lunar city of 10,000 people by the middle of the next century and is busy designing resort hotels, spaceship landing sites and roads — not only for the city itself, but for tourist trips to scenic craters and mountain ranges, and perhaps to the historic Apollo sites such as Apollo 11's Tranquility Base.

If international partners agree to build a moon base, they could develop many techniques there that could enable them to again join hands to mount an expedition to Mars. During years in an Antarctica-type lunar base, researchers could gain experience on all aspects of living on a distant planet, using local resources for building structures and growing food, and designing closed-loop life support systems.

The tent-like Lunar Surface Vehicle Servicing Facility (depicted in top picture) provides thermal protection for the surface transport vehicles and the several work modules that are used in conjunction with the transporters. The work modules include drilling rigs and cranes to support telerobotic activities, as well as pressurized work modules to support far-ranging human exploration of the surface (bottom picture). The moon "garage" is constructed with multilayer insulation blankets to form a protected work space for the suited astronauts and the equipment. Artwork by Pat Rawlings/SAIC.

In this lunar outpost concept, a lunar oxygen production plant, set between two large solar panels, is generating a supply of rocket fuel that will be used for later journeys to Mars.

NASA's "Dante" robot, here making a test climb up a slag pile at Carnegie Mellon University, will rappel itself down the sheer crater of Mount Erebus in the Antarctic to get scientific data that could help pave the way for the exploration of Mars.

Using a moon base or a space station as a jumping off place for Mars has been considered, but most engineers say it would be easier and cheaper to build a mammoth new rocket for a direct flight from Earth. NASA is working with aerospace companies to develop such a rocket, called the X-33, to replace its aging space shuttle fleet. The agency wants the vehicle to be operational by 2004.

Russia's giant Energia rocket also is considered a possibility for launching a manned mission to Mars.

As nations ponder such a trip, an armada of unmanned spacecraft is being readied to probe secrets of the Red Planet in the next decade. As many as 28 flights are planned, many of them multi-national efforts.

In concert with Russia, Japan and the European Space Agency, NASA plans to launch at least two missions every time Earth and Mars are aligned properly for such flights. That occurs every 26 months, with each opportunity lasting 30 days. Each multi-million-mile outward journey takes about eight months.

The first to be launched, from Cape Canaveral in November 1996, will be the *Mars Global Explorer*, which is to orbit Mars. A month later, the *Pathfinder* will set forth to land near a dried-out Martian riverbed and dispatch a small rover to gather photos and scientific information about the area.

The Russians, French and Japanese all are developing Martian robots. The assault on the planet will culminate in 2003 and 2005 with international missions that will include seven landers, two missile-like penetrators and a communications satellite to relay data to Earth from Mars.

What these probes learn, and how well nations work together will help determine whether humans from many lands will be sent to Mars as representatives of Planet Earth.

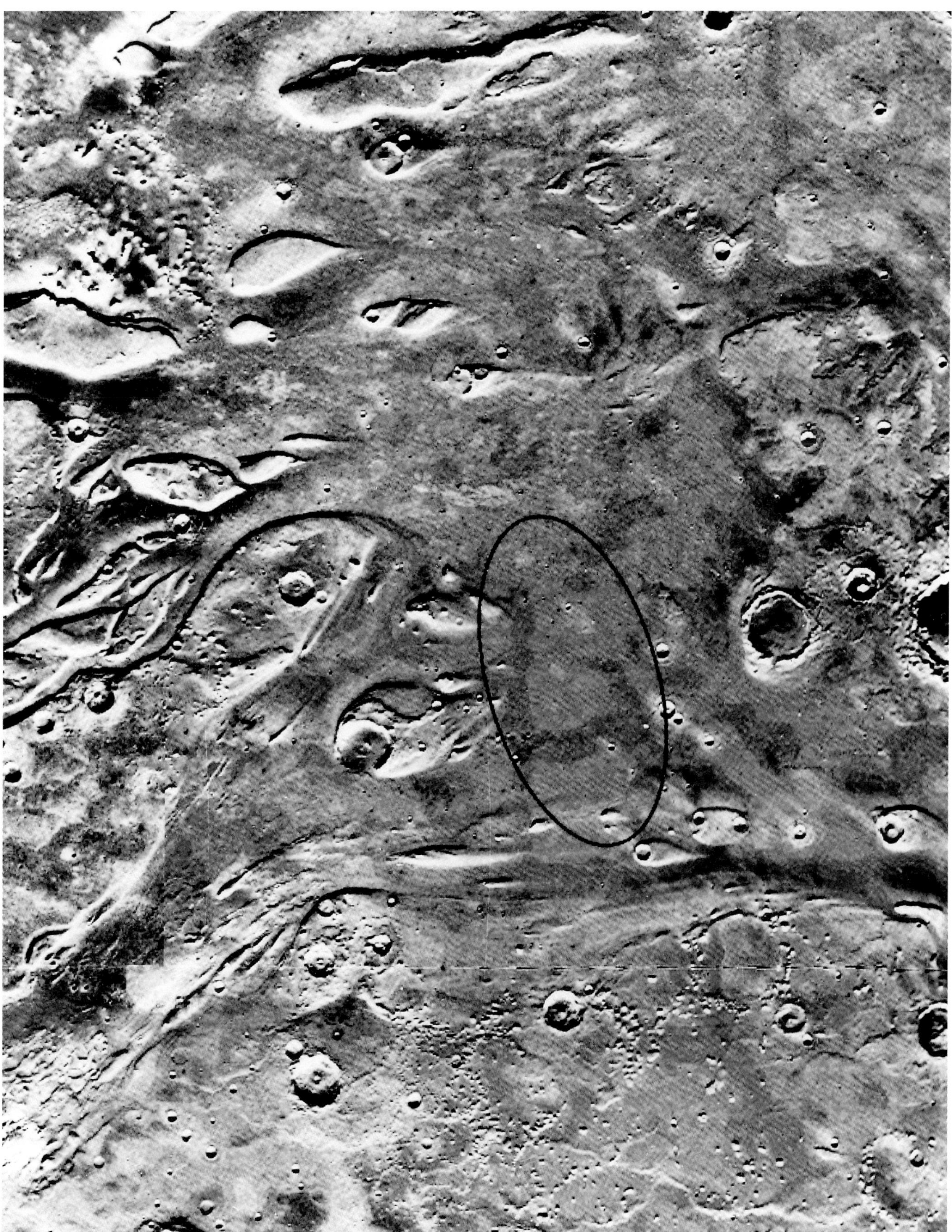

The *Mars Pathfinder* landing site, Ares Vallis, is in the Chryse Planitia region in the northern hemisphere of Mars. It is near the site explored by the *Viking 1* lander in 1977. This mosaic, made from *Viking* orbiter data, shows the large outflow channels that emptied into Chryse Planitia. Ares Vallis formed from the release of water from the Martian subsurface and flowed across the surface, creating the channels and the large islands. The landing site,a smooth depositional surface where the flood waters deposited the sediments carved from the channel, should provide a wide variety of rock types for analysis.

In this artist's concept, the *Mars Pathfinder* lander has landed and opened its petals. It transmits to Earth key data collected during the entry, descent and landing phases on the performance of the lander systems and the surrounding environment. The open petals are covered with solar cells to provide electrical power for the surface operations and survival of the lander in the Martian environment. A gold thermal blanket protects the computer, transmitter and other lander electronics from the temperature extremes typical on Mars.

The lander camera photographs the surrounding terrain. After the pictures are examined, commands from Earth direct the microrover to drive off the lander petal onto the Martian surface. It then becomes a free-roving vehicle with its own camera system for navigation and science investigations, reliant on the *Pathfinder* lander only for communications to and from Earth. The microrover conducts technology experiments important to the design of future Mars roving vehicles. It also places a spectrometer against Martian rocks and surface materials to determine their elemental composition.

Why Mars? Because it is the only other planet in the solar system that could sustain life. Venus is a blistering hot world where 900-degree heat is trapped in a thick atmosphere of carbon dioxide laden with sulfuric acid clouds. Mercury is a parched and cratered wasteland. Saturn's atmosphere of helium and hydrogen is alive with monster storms and winds up to 1,100 miles an hour. In addition to being uninhabitable, Neptune, Uranus and Pluto are too far away.

Mars may appear to be a barren place, but beneath its sands are oceans of water in the form of permafrost. Its atmosphere is mostly carbon dioxide, which could provide large supplies of the two most important biological elements in a chemical form for plant life. Also present are nitrogen, silicon, sulfur, phosphorus, inert gases, all the metals and other raw materials needed not only to sustain life but also to create an advanced technological society.

Once humans reach Mars, facilities could be established to support both exploration and development of techniques for converting the planet's raw materials into food, fuel, propellant, plastics, metals and other necessities for building an ever-expanding population.

Once the production infrastructure is in place, as many as 100 colonists a year, at first, could migrate to Mars, producing a population growth on Mars in the 21st century comparable to that which occurred in the new world of America in the 17th.

Researchers believe that colonizing Mars will drive the development of new and more powerful sources of energy and ever faster, safer means of space transportation. These in turn would enable humans to travel ever deeper into the solar system to inhabit other worlds and learn of the wonders that exist there.

Whether the United States and other nations have the will to commit to such exploration is unknown at this time. But if they do, they will have to work together as representatives of all mankind. For no single nation has the resources to do it alone.

If international cooperation on the space station can lead to more ambitious undertakings, humans indeed can follow their dreams to distant stars.

This NASA Hubble Space Telescope image of the center of globular star cluster M15 reveals a new population of about 15 very hot and blue stars isolated at the core. M15 is located 30,000 light-years away in the constellation Pegasus.

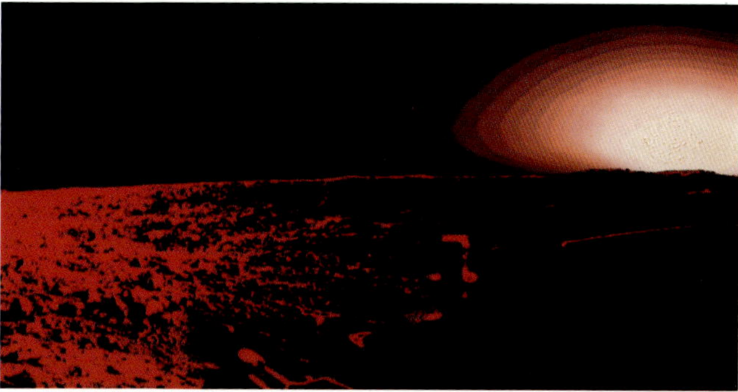

Why Mars? It is the only planet in the solar system that can sustain life. This Mars sunset was captured by the *Viking* lander.

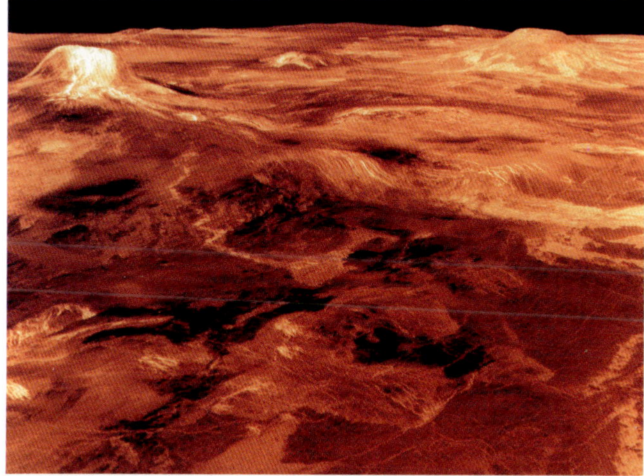

Venus is a blistering hot world, as this *Magellan* picture proves.

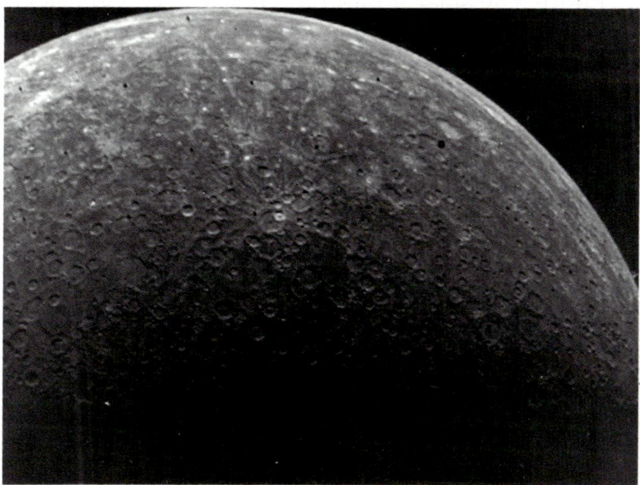

Mercury, as seen by *Mariner 10,* is a parched and cratered wasteland.

Saturn has monster storms and 1,100 mph winds. This image was taken by Hubble in December 1994.

Rockwell

From the first Americans in orbit to the first humans on the moon to today's routine operations of the space shuttle, the programs that define the Space Age have relied on Rockwell systems. Now, America and the world are ready to take the next logical step in exploring the cosmos — the construction of a permanent manned space station. Once again, Rockwell, a leader in a variety of high technology arenas, will be there.

Powering the Station

The international space station, a joint venture of the U.S., Russia, Japan, Canada and the European Space Agency, is scheduled to begin assembly in 1997, with completion set for 2002. Rockwell, continuing its role as a leader in space, is providing the station's electric power generation and distribution system. The end-to-end electric power system architecture will provide all user and housekeeping electricity, taking power from the station's solar arrays and distributing it to onboard equipment and experiments. As the space station is assembled, the Rockwell-designed system will expand to ultimately provide 110 kilowatts of continuous power, equal to the amount of electricity required to power 40 homes on Earth. The solar arrays on the space station will be two-thirds the length of a football field.

Getting There Via Shuttle

Ferrying astronauts, equipment and supplies to the space station — first to build it, then to keep it functioning — will be the Rockwell-built space shuttle, the world's first reusable space transportation system.

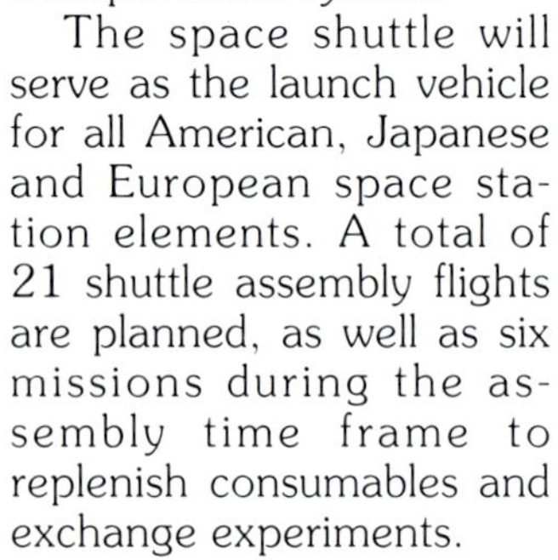

Space shuttle *Discovery* lifts off in September 1994.

The space shuttle will serve as the launch vehicle for all American, Japanese and European space station elements. A total of 21 shuttle assembly flights are planned, as well as six missions during the assembly time frame to replenish consumables and exchange experiments.

Rockwell is currently modifying America's fleet of space shuttles to improve performance, safety and reliability, reduce the cost of operation, and keep technologies up to date. To increase the fleet's payload capacity, the overall shuttle system will be adapted to accommodate an additional 15,000 pounds. Shuttle cockpit displays and controls are being upgraded with a multifunction electronic display subsystem ("glass cockpit"). Use of the Global Positioning System (GPS), a constellation of Rockwell-built satellites providing precise, three-dimensional navigation information, also is expected to be incorporated into the shuttle fleet.

Rockwell is providing the space station's electric power generation and distribution system, which ultimately will provide 110 kilowatts of continuous power.

The shuttle system is capable of adapting to meet the emerging space requirements of the late 1990s and beyond. Its exceptional design flexibility allows for such evolutionary upgrades as a long-duration (30- to 90-day) flight capability; liquid-fueled rocket boosters; an automated orbiter kit that would allow orbiters to fly without a crew as mission needs dictate (thereby accommodating up to 12,500 pounds of additional payload) and even a reusable cargo vehicle — an orbiter without a crew.

With a design life of at least 100 missions per vehicle, evolution of the space shuttle system provides the U.S. with a viable, near-term, low-risk alternative to developing a new launch system. The shuttle can also serve as an invaluable test bed for the enabling technologies that will be required to develop America's next-generation launch system.

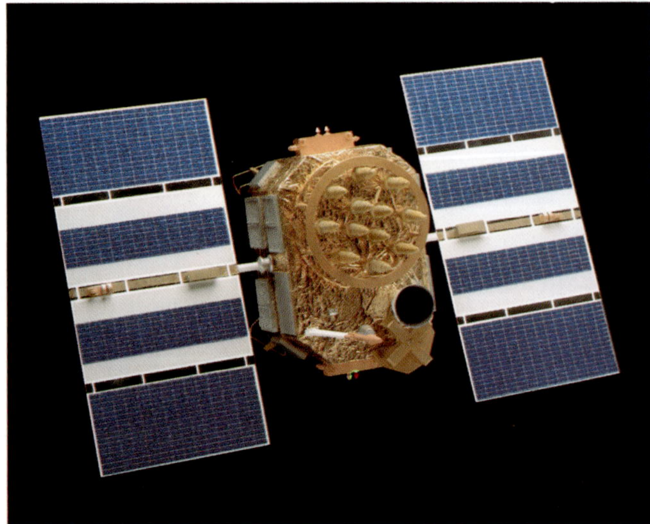

Rockwell developed and produced the network of satellites for the Global Positioning System.

International Cooperation

Among the modifications planned for the space shuttle *Atlantis* is the addition of hardware that will enable it to dock with Russia's space station, *Mir*. Beginning in summer 1995, *Atlantis* will perform a series of docking missions with *Mir* in a precursor phase of the space station program that will prove operational concepts and gather essential life science data. These missions will also establish a cooperative framework for joint U.S.-Russian space station assembly and operations.

Modifications on *Atlantis* were completed as part of the vehicle's scheduled orbiter maintenance down period at Rockwell's Orbiter Modification Facility in Palmdale, California, from October 1992 through May 1994. The *Atlantis* effort was completed two weeks ahead of schedule and approximately $2 million under budget. NASA cited Rockwell's "team-based organizational approach, with its emphasis on high-performance teamwork and 100 percent workmanship quality" as the driving force behind this success.

Space shuttles *Endeavour*, *Discovery* and *Atlantis* will eventually be modified to enable docking with the international space station. Under a NASA contract, Rockwell's Space Systems Division began work in March 1994 on two additional external airlocks, associated equipment and orbiter modifications for shuttle docking with the space station. A crew lock will support spacewalks during space station maintenance.

The Rockwell-built orbiter docking system that will be used when space shuttle *Atlantis* and Russia's space station *Mir* link up in 1995 incorporates a Russian-built docking mechanism.

From the Beginning

A world leader in spacecraft and rocket propulsion systems, Rockwell's aerospace business has been involved in every manned American space program — and many of NASA's and the Department of Defense's most prominent unmanned efforts. Rockwell was instrumental in many of America's early space successes, developing technologies in space structure, guidance and power. The X-15 research vehicle; propulsion systems for Jupiter, Thor, Redstone, Atlas, Delta and Saturn boosters; and the Apollo spacecraft and Saturn V S-II stage are all Rockwell achievements.

Rockwell's march toward space began in the 1950s with the development of the rocket-powered X-15, the only aircraft to reach an altitude of 67 miles — the edge of space. The journey continued with the production of liquid-propellant rocket engines, such as the Redstone booster, which put America's first satellite in space. That was followed by propulsion systems for Atlas/Mercury. In 1961, after President John F. Kennedy committed the nation to a manned lunar landing before decade's end, NASA awarded Rockwell the most demanding project of all: Apollo.

The rocket-powered X-15, built by Rockwell, was the first winged vehicle to reach an altitude of 67 miles — the edge of space.

Rockwell's Apollo Role

Rockwell's Space Systems Division (SSD) in Downey, California, was the principal contractor to NASA for the Apollo command and service modules, the launch escape system, the spacecraft-lunar module adapter, and the Saturn V second stage (S-II).

Rockwell was the predominant supplier of the propulsion systems for the massive Saturn V launch vehicle and was the only major contractor with hardware in almost every segment of the Apollo-Saturn stack (the Saturn launch vehicle and Apollo spacecraft). The stack stood 363 feet high, weighed 6.4 million pounds and generated some 7.5 million pounds of thrust at liftoff.

Teams from Rockwell performed the engineering design, manufacturing and exhaustive pre-flight testing of the Apollo command and service modules.

Another major achievement was the design, development and construction of the huge second stage of the Saturn V vehicle — 33 feet in diameter and 81.5 feet in height — produced by Rockwell in Seal Beach, California.

The spacecraft lunar module adapter, a thimble-shaped structure 28 feet tall and 22 feet in diameter at its base, truncated to 13 feet in diameter at its mating point with the service module, was constructed at the company's Tulsa, Oklahoma, facility. The Tulsa facility also produced the primary aerostructure for the Apollo service module and the Saturn V instrument unit.

All of the primary Saturn launch vehicle engines were built by Rockwell's Rocketdyne Division in Canoga Park, California, including the ascent engine of the lunar module, which boosted the astronauts

Left: Astronaut Buzz Aldrin prepares to step onto the moon during Apollo's first lunar landing in 1969. Right: American Tom Stafford and Russian Alexei Leonov work together during the first international manned space mission in 1975.

from the moon's surface back to the lunar-orbiting *Columbia* command module.

Collins divisions in Cedar Rapids, Iowa, and Richardson, Texas, provided the communication and data system installed aboard the Apollo command module. Their equipment also transmitted live television signals from the moon, including astronaut Neil Armstrong's first video and voice messages.

Back on Earth, ground stations employing Collins' 30-foot diameter antennas monitored Apollo as it orbited the Earth. Three 85-foot-diameter antennas bridged the 239,000-mile distance from the moon to Earth.

The Autonetics business in Anaheim, California, developed the entry monitoring system and sequential events control system on the spacecraft, and provided electronic equipment for both the Apollo and Saturn vehicles.

The North American Aircraft Division, now located in Seal Beach, California, built the Apollo's probe and drogue assemblies used in docking the command and lunar modules, as well as the launch escape tower and boost protective cover.

Products from Allen-Bradley (A-B), an industrial automation leader based in Milwaukee, also contributed to the success of the Apollo missions. A-B's carbon composition resistors were "basic building blocks" for possibly every circuit board installed in the lunar module and other equipment.

In all, there were 15 manned and unmanned Apollo missions, and six lunar landings. Rockwell, then known as North American Rockwell Corporation, held a dominant role in each, as well as with the Skylab and Apollo/Soyuz missions that followed.

In 1994, America celebrated the 25th anniversary of Apollo 11 and the "one small step for a man, one giant leap for mankind" made on July 20, 1969, when Neil Armstrong placed his foot on the moon's surface. Armstrong and crewmate Edwin "Buzz" Aldrin spent 21 hours and 36 minutes on the moon, including a combined total of four hours and 22 minutes walking on the surface, setting out experiments and collecting 49 pounds of rock and lunar soil.

"The trip to the moon was a race that every American wanted us to win," Donald R. Beall, chairman and chief executive officer of Rockwell stated on the occasion of the 25th anniversary of the moon landing. "It wasn't just a political battle. Nor was it the test of a few heroic astronauts. It was the pioneering spirit shared by thousands of scientists, engineers and workers who joined together in the greatest intellectual exploration known to mankind."

Space Shuttle Continues the Quest

After Apollo, the ongoing exploration and conquest of space continued with the space shuttle. The orbiter *Columbia* was launched on the first shuttle mission on April 12, 1981. Rockwell designed, developed and constructed the space shuttle orbiters and their main engines, and has an ongoing role in orbiter maintenance, upgrades and modifications. The company is responsible for shuttle cargo and system integration and other operation support services, including mission support and orbiter and main engine logistics.

The Space Systems Division is the prime contractor to NASA for the space shuttle program. The Rocketdyne Division produces the space shuttle's main engines. Collins Avionics & Communications Division supplies the shuttle's TCN-252 TACAN, a tactical air navigation system. The space shuttle's main engine, the only reusable liquid rocket engine ever built, operates at greater temperature extremes than any mechanical system in common use today. The fuel, liquid hydrogen, is minus 423 degrees Fahrenheit, the second-coldest liquid on Earth. When burned with liquid oxygen, the temperature in the engine's combustion chamber reaches 6,000 degrees Fahrenheit.

The three engines on the shuttle orbiter provide the equivalent of just over 37 million horsepower, the majority of the total specific impulse required to obtain near orbital velocity. Although the shuttle travels from zero to 60 miles per hour in a time not much faster than the average sports car (approximately five seconds), it reaches the speed of an average commercial jetliner (about 550 mph) in approximately 34 seconds. It exceeds the speed of the fastest airplane (about 1,850 mph) in about one and a half minutes, and attains orbital velocity (approximately 17,231 mph) in about eight and a half minutes. Landing speeds vary from 213 to 226 miles per hour.

Columbia, which was launched on the first shuttle mission in 1981, completes a microgravity research mission in July 1994.

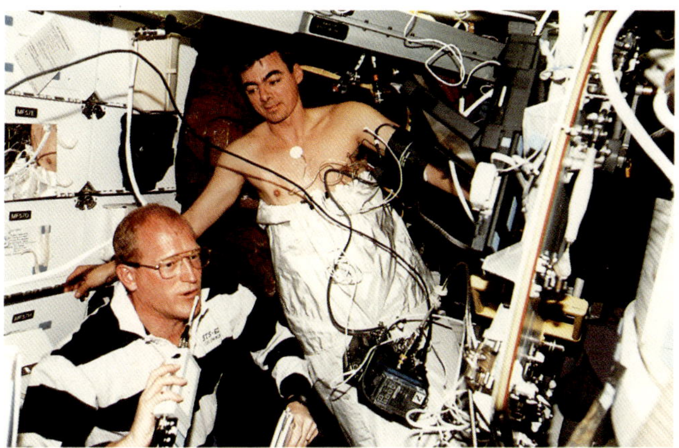

Astronaut Charles Gemar talks to ground controllers while assisting Andrew Allen in an experiment in which he "soaks" in a lower body negative pressure apparatus during a March 1994 mission aboard *Columbia.*

During *Discovery's* September 1994 mission, astronaut Mark C. Lee tests a new extravehicular activity rescue device.

The reusable thermal protection system developed by Rockwell for the space shuttle shields the structural skin of the orbiter from temperatures ranging from minus 250 degrees Fahrenheit in the cold soak of space, to entry temperatures that reach nearly 3,000 degrees Fahrenheit.

Through 1994, the shuttle orbiters had flown more than 7,800 orbits around the Earth, and traveled more than 200 million statute miles. Through more than 65 successful missions to date, the space shuttle orbiter fleet has deployed more than 50 satellites, including interplanetary spacecraft and orbiting observatories; repaired and/or retrieved malfunctioning spacecraft payloads that could not have achieved mission successes through other means; and flown more than a dozen Spacelab missions.

The hundreds of experiments conducted aboard the space shuttle have investigated life sciences, materials sciences, combustion science, solar science and physics, space plasma physics, atmospheric studies, biotechnology, Earth observations, astronomy and the study of the behavior of metals, semiconductors, bioprocessing and fluids.

Peter J.K. Wisoff stands on a mobile foot restraint aboard *Endeavour* as he practices procedures that were later used in repairing the Hubble Space Telescope.

The space shuttle has proven to be far more than just America's space transportation system. It has been a proving ground for the space systems and programs of tomorrow.

Future Technology

Although the space shuttle is expected to satisfy America's manned space transportation needs for the near future, Rockwell is at work developing key technologies that will be incorporated into a future reusable launch vehicle that will allow economical, reliable access to space in the 21st century.

Under cooperative research agreements with NASA's Marshall Space Flight Center, Rockwell will design, fabricate and test long-lead technologies such as reusable cryogenic propellant tank systems, lightweight, durable thermal protection systems, composite primary structures and main propulsion systems.

Beyond space transportation, Rockwell is developing new space surveillance and anti-missile defense systems for the Department of Defense, including the Exoatmospheric Kill Vehicle (EKV), an Army missile defense program; Brilliant Eyes, a ballistic missile early warning satellite defense system; the Air Force P91-1 Advanced Research and Global Observation Satellite Program (ARGOS), which will perform global observations of the ionosphere from four sensors simultaneously, as well as technology demonstrations; and the Mars Surveyor Program (MSP), a small generic spacecraft/lander designed to survey the surface of Mars.

A Multi-Faceted Company

While Rockwell is well known for its successes in aerospace, it is also a diversified, high technology company holding leadership market positions in automation, avionics, defense electronics, telecommunications, automotive components and graphic systems. The multi-faceted company has annual worldwide sales of more than $12 billion.

Instron Corporation

The Global Standard for Materials Testing Equipment

Since the dawn of time, humans have sought ever better materials to construct shelters, provide safety and food, and make tools. More recently that quest has expanded to include the search for materials to build a safe home in space and vehicles to travel there. Determining the strength and durability of materials to withstand the hostile environment of space are test instruments and systems built by Instron Corporation of Canton, Massachusetts.

For nearly 50 years, material scientists around the world have used test systems designed and built by Instron to evaluate virtually all of the earth's natural and man-made materials. When NASA needed a test system to evaluate materials for the space environment, the agency turned to the company that has set the international standard for precision, performance and reliability of testing systems: Instron.

Today, the NASA Biaxial Servohydraulic Testing System, custom-designed by Instron, is evaluating advanced composites at high temperature both in air and vacuum. The system's computer work station runs a LabVIEW program, specially designed by Instron software engineering to simulate service conditions being modeled.

Instron 8580 four-axis test system with LabVIEW Software Drivers.

Dr. Tuncer Cebeci, Chair, Aerospace Engineering Department, California State University, Long Beach.

The Aerospace Engineering Department of California State University, Long Beach, (CSULB), under the direction of Dr. Tuncer Cebeci, turned to Instron for systems to evaluate new materials for NASA's High Speed Civil Transport Research (HSCT) program. CSULB is engaged in the program to evaluate these advanced composites in conjunction with McDonnell Douglas Corporation (MDC). Preliminary studies at McDonnell Douglas indicated that currently available aerospace materials would not satisfy high speed civil transport requirements.

The CSULB Composite Durability Laboratory is evaluating potential HSCT materials over long terms and extended temperatures. The lab has ten all-digital,

California State University, Long Beach, Composite Durability Laboratory.

Instron 8500 Series servohydraulic test systems, eight equipped with specially designed -65°F to +600°F temperature chambers and two equipped for additional research. All systems are data compatible with the 27 Instron systems located at the NASA-LaRC laboratory. Instron application specialists worked closely with CSULB, MDC and NASA to provide the systems and software the HSCT program requires. As Dr. Cebeci, chair of the CSULB Aerospace Engineering Department, explained, "Only Instron gives us the compatibility this NASA program requires."

Instron instruments have long been used to evaluate materials for aerospace applications. They performed developmental tests on space shuttle's thermal protection system components, including the LT-900 tiles, adhesive and fiber/silicone FRISI insulation. They tested the structural integrity of materials used in rocket nozzles and solid rocket propellants. And, they continue to measure the ability of materials and composites to endure 2900°F temperatures for use in future spaceships capable of traveling at hypersonic speeds.

In addition to evaluating materials for the aircraft and aerospace industries, Instron test instruments are used worldwide to evaluate the strength-related properties of materials used in such diverse fields as nuclear power generation, biotechnology, food technology, medicine, transportation and defense.

From Textiles to Composites

Instron Corporation was founded by Harold Hindman and George Burr, two MIT researchers who developed the tensile tester and brought it to market in 1946. Originally developed for use with textile materials, their innovative test instrument, which pioneered the application of electronic load weighing and servo control techniques, quickly proved useful to all the other industries that would enter the race for new and improved materials.

Materials scientists and engineers, concerned with all aspects of the performance of materials, accelerated the search for lighter, stronger, more energy efficient materials. The search spawned new initiatives into composites, the marriage of high strength fibers and a matrix of plastics or metals. In order to evaluate the effectiveness of a new material, alloy or composite, and the design of the components and structures that used them, materials scientists and engineers needed accurate, reliable instrumentation capable of providing repeatable results anywhere in the world. Instron has supplied instrumentation for use in all phases of materials testing, from research through production and quality control.

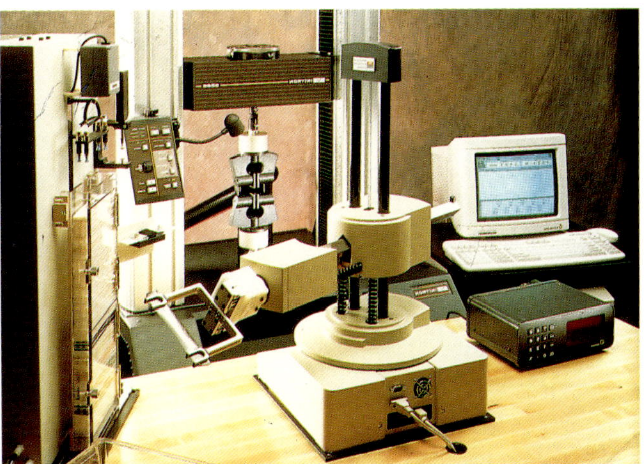

TestMaster™ Automation System performs unattended, round-the-clock testing.

Instron's comprehensive line of materials testing products include electromechanical instruments, servohydraulic instruments, structural component systems, computer software, temperature chambers, rheological instruments, hardness and spring testers. The company has more than 40,000 equipment installations worldwide, used by industry, governments and educational institutions. Behind every instrument and system developed at Instron are specialists with skills ranging from mechanical and electronic engineering to the writing of software programs that meet the special requirements of material testing. This capability is largely responsible for Instron's continued worldwide leadership in materials testing.

Series 5500 Universal Test System.

A New Generation of Materials Testing Systems

Instron developed a new generation of materials testing equipment that uses personal computers for complete test set-up, control and data analysis. These digitally-powered instruments perform the full range of tensile, compression, flexure, shear, torsion, and cyclic tests. With the click of a mouse, the operator can pull the appropriate test method from memory and instantly establish test speed, display criteria, results calculations, statistical data, and other report information. Built-in flexibility allows the operator to quickly change over from a tensile test of a delicate fiber material to a flexural test of the completed composite. Computer controls allow testing to be synchronized with a high/low temperature chamber that provides conditions that simulate an intended application, for measuring properties over thermal extremes.

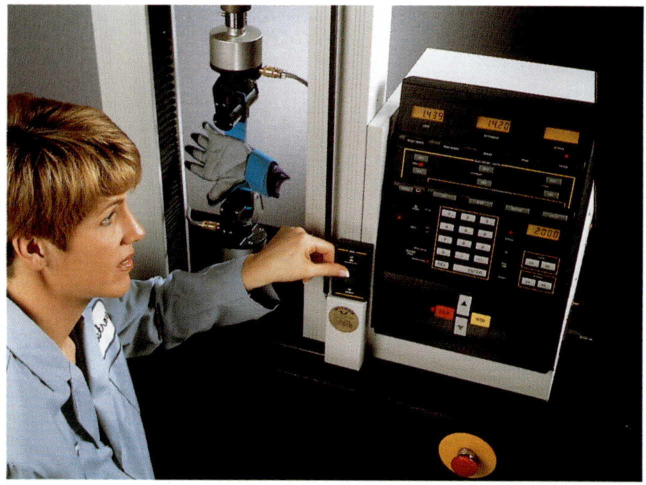
Instron Model 4411 testing a sport glove for quality.

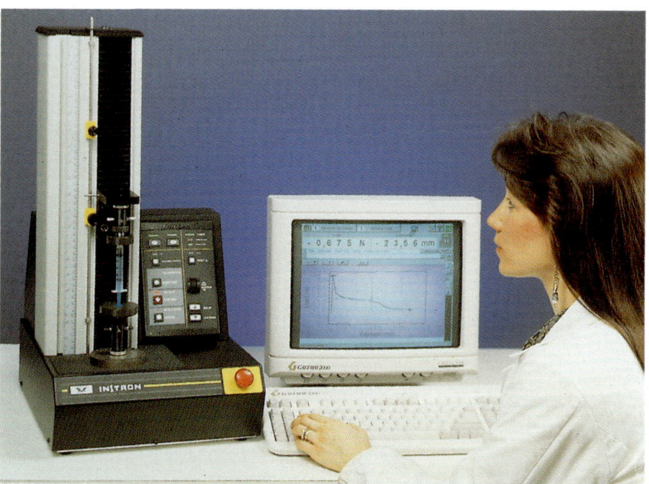

Instron Mini 55 testing system performing a biomedical test.

Instron Products

The new generation, computer-based Series 5500 is capable of testing all kinds of specimens: plastics, composite, metals, elastomers, textiles, fabrics, finished products and more. It measures the properties of materials in tension, compression, peel, shear, and flexure.

The user interface incorporates both a control panel and a screen display that makes setting up and running tests quick and easy. The personal computer, with Merlin Windows-based software, is used to set up test procedures, collect data, calculate test results and generate reports. The test is run from the control panel on the load frame, which has all the test functions needed — starting, stopping, moving the crosshead, calibrating, resetting the gauge length. Series 5500 does all the statistical calculations that quality control and research laboratories need.

Model 8516 fatigue test system.

The Instron Mini Testers, the Mini 55, part of the Series 5500 family, and the Mini 44, provide high performance materials testing in a compact size. Used by research and university labs, and quality control environments, they are ideal for testing biomaterials, film, fibers, textiles, fine wire, food or packaging materials.

Instron Series 4400 materials testing instruments provide a broad range of test capabilities for performing tensile, compression, shear, peel and flexural testing of most materials and components, including textiles, metals, fasteners, wire, insulation, elastomers and rubber. The operator panel allows instantaneous test set-ups or changes, and load frames can accommodate from 450 to 135,000 pounds.

Typical test parameters are automatically computed, including tensile strength, elongation, peak and break load and strain. Other results, such as modulus of elasticity or flex modulus, are available using Instron's optional Series IX software.

The 8500 Plus Servohydraulic Testing Instruments are designed to fulfill advanced materials testing requirements into the future. They are ideal for fatigue testing of metals, composites, and components, and are used in various fields including aerospace, automotive, specialty materials, engineered materials, biomedical, and university research.

The 8500 Plus provides digital test control, data acquisition and offers a wide range of test frequencies with environmental and temperature control. The control panel gives the operator simple access to full closed-loop control of load/stress, position or strain. The system incorporates an industry standard interface to a personal computer and a diverse range of Windows-based software is available for many applications. A range of accurately aligned load frames enable dynamic loads from 5,000 to 500,000 pounds. Instron offers a Retrofit Package to up-grade older systems to high performance digital control.

Instron's new 8516 hydraulic testing system is designed to give maximum return on investment, through improved productivity in a value-engineered package. It is ideal for high cycle fatigue, tensile/-compressive, fracture mechanics, low cycle fatigue, flexure, components and environmental testing.

Software, Accessories, Service and Support

Instron software is designed, developed and rigorously tested by experts in fields ranging from textiles to fatigue crack growth and service life simulation.

Series IX is a full-featured software package combining all the power of the world's most popular materials testing software and all the ease of Microsoft® Windows®. Tensile, compressive, flexural, peel, tear and friction test methods are available in a multi-application package for data acquisition, control and analysis of monotonic testing.

Fast Track™ is Instron's family of flexible, Windows-based software for servohydraulic testing applications, which includes:

- MAX – A general fatigue program for most fatigue applications, including service simulation and multi-axis testing.
- Applications Programs – Fracture mechanics, low cycle fatigue, spectrum loading, service simulation, elastomer testing, soils and asphalts, tensile, comparison and general monotonic tests.
- Custom program with LabVIEW® – Instron offers drivers so you can write your own applications or Instron can write them for you.

The broad range capabilities of Instron's materials testing systems are further enhanced with a wide selection of grips, fixtures, and accessories that can be added to the basic system to meet specific application requirements. Included are extensometers (strain gauge, elastomeric, high resolution digital, and patented non-contacting video); temperature cabinets; special grips, special purpose fixtures and load cells.

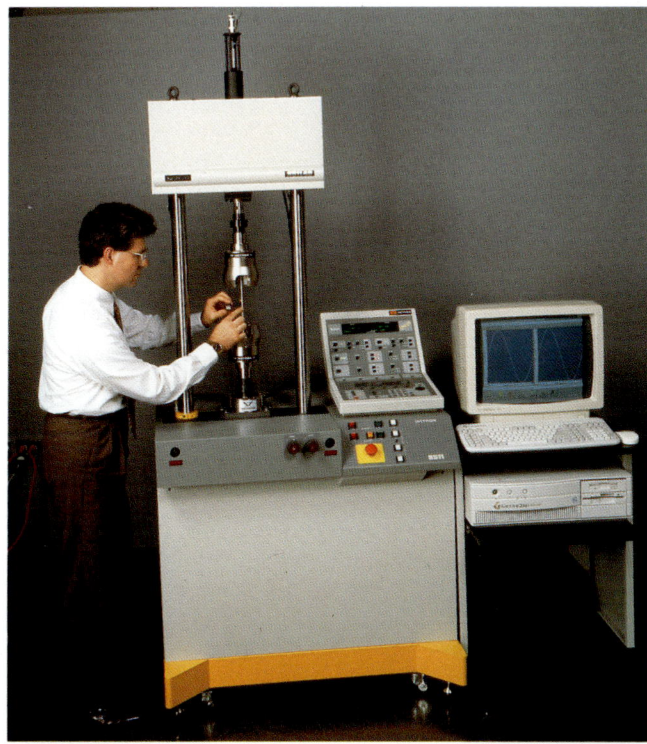

FastTrack™, Instron's Windows® family of software, features a range of easy to use Windows-based fatigue testing software.

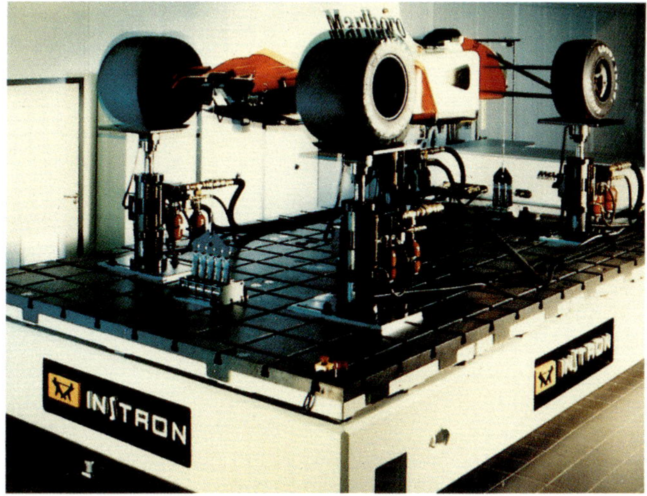

The McClaren racing car uses advanced composites, and is being tested in Instron's structural test system simulating Formula 1 racing stresses.

Full Service Around the Globe

Instron factory trained employees staff the company's offices worldwide, and are available for on-site training, maintenance and repair. Training courses also are provided at the Canton, Massachusetts and High Wycombe, England factories.

An international corporation with more than 1,000 employees and annual sales exceeding $136 million, Instron has major manufacturing facilities in the United States and England. Sales and service facilities are located in 12 U.S. locations and 15 foreign countries. The international scope of Instron's operations, and its comprehensive product lines, make it the only full service materials testing company serving the entire world.

Instron's corporate family includes Wilson Instruments, which manufactures a complete line of test systems including microhardness, Brinell, and the Rockwell® hardness tester developed more than 70 years ago and still a worldwide standard; Severn Furnaces Limited, a world leader in the field of radiant furnace technology; Laboratory MicroSystems, Inc., the leader in PC based Laboratory Information Management Software; Instron-Wolpert GmbH, a manufacturer and distributor of hardness, spring and impact testing equipment worldwide, and Shore Instruments, manufacturer of durometers for measuring rubber hardness.

Instron and Fast Track are trademarks of Instron Corporation. Windows is a trademark of Microsoft Corp. LabVIEW is a trademark of National Instruments. Wilson and Rockwell are trademarks of Wilson Instruments, Division of Instron Corp.

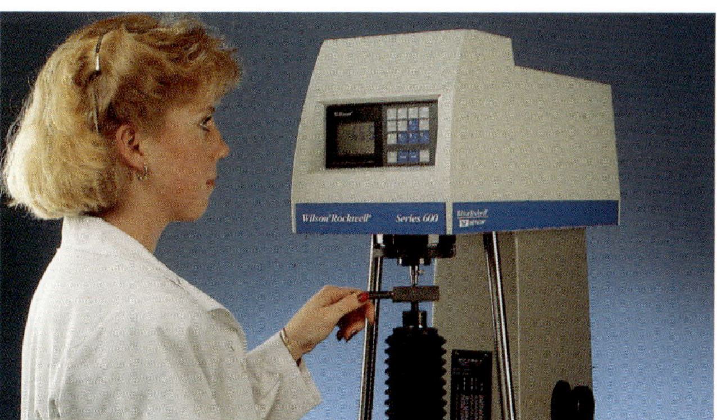

New Wilson® 600 Series Rockwell® tester.

Hughes Electronics

When the Apollo astronauts landed on the moon, Hughes-built spacecraft were there to greet them. In the mid-1960s, the Surveyor spacecraft were sent to the moon to scout potential landing sites for the astronauts.

The five Surveyors that soft-landed on the moon not only proved the technology of a lunar landing spacecraft (soft landing, motor restart and controlled movement across the moon's landscape), but also conducted scientific experiments and sent back color pictures of a solar corona, the lunar surface and planet Earth. Portions of Surveyor 3, returned to Earth by Apollo 12 astronauts after sitting on the moon's Ocean of Storms for more than two and a half years, provided a baseline against which the environmental effects of lunar existence could be measured. They also have aided in the selection of materials, design components and equipment for long-term manned and unmanned space exploration systems.

In addition to landing on the moon, Hughes worked with NASA to build the first synchronous orbit communications satellite, Syncom. In the years following, Hughes spacecraft and instruments have descended into the atmospheres of Venus and Jupiter and studied the sun. They have pioneered international voice, video and data communications, helped meteorologists forecast the weather and aided in the study of the stars and of Earth's resources.

Surveyor 3 is visited by Apollo 12 astronauts. Lunar environmental effects on parts returned to Earth aided in the development of materials for long-term space exploration.

Hughes Synergy

Much of Hughes' space work resides in its Hughes Space and Communications Company (HSC) business unit, the 1991 winner of NASA Goddard Space Flight Center's award for excellence in quality and productivity. HSC is the world's leading commercial communications satellite manufacturer, a major supplier of scientific instruments to NASA, and a developer and integrator of space and ground systems for customers worldwide.

Santa Barbara Research Center (SBRC), another business unit, also is a major player, with its development of sensors used on spacecraft orbiting Earth and other planets. SBRC has designed and built a wide variety of instruments, such as multispectral imagers, radiometers, spectrometers, polarimeters, and sounders used for meteorological applications, planetary exploration, acquisition of Earth-resource data, and as star trackers. Every successfully launched SBRC instrument has operated properly at turn-on, and their combined operating life now exceeds 250 years.

Hughes Danbury Optical Systems, Inc. (HDOS) joined Hughes as a wholly owned subsidiary in 1989, bringing with it more than 30 years of experience in building scientific instruments for NASA. HDOS is a major supplier of large optics for astronomy and space science programs, among them NASA's Hubble Space Telescope and Advanced X-ray Astrophysics Facility (AXAF). HDOS's Optical Telescope Assembly and three Fine Guidance Sensors have enabled Hubble to be an extremely powerful telescope capable of producing images of many stellar objects with far greater clar-

ity and detail than ground-based observatories. HDOS fabricated four nested, cylindrical, grazing incidence mirror pairs for AXAF, which is scheduled for launch in 1998.

Hughes Information Technology Corporation (HITC) works with NASA to develop premier information technology systems, from programs such as Mission to Planet Earth to the exploration of Jupiter. HITC designs and builds satellite ground systems; writes software to manage complex operational tasks; designs, integrates, operates and maintains complex data networks; and brings fresh, innovative and cost-effective approaches to business and data management problems. When the Shoemaker-Levy-9 (SL-9) comet slammed into Jupiter in the summer of 1994, HITC staff supported the National Space Science Data Center at Goddard with a World Wide Web server that made SL-9 images from a variety of sources available in near real-time. NSSDC logged more than 9,000 different users with 360,000 accesses to the SL-9 images alone via the server.

Mission to Planet Earth

Charged along with several other government agencies with developing a detailed understanding of Earth system processes and their contribution to global climate change, NASA has turned its attention to home base with Mission to Planet Earth. Studies of the Earth's air, water, land, living matter and their interactions are made possible with the Earth Observing System (EOS), a series of satellites carrying advanced remote-sensing instruments, and a variety of other mission elements, including Earth Probes and operational and research satellites. Hughes is playing a major role in Mission to Planet Earth by providing the following:

- Two components aboard the Tropical Rainfall Measurement Mission (TRMM) observatory: HSC's TRMM Microwave Imager (TMI) and SBRC's Visible and Infrared Scanner (VIRS). A shared mission with Japan and one of the series of EOS Earth Probes expected to fly in late 1997, TRMM is designed to advance understanding of rainfall, heat, and other energy exchange processes, and to determine the rate of rainfall and total rainfall that occurs over the tropics and subtropics. VIRS and TMI are the primary components of the TRMM rain package that will aid in meteorology and climate research.
- SBRC's Sea-Viewing Wide-Field-of-View Sensor (SeaWiFS). Scheduled for launch as an EOS Earth Probe early in 1995, the sensor will provide ocean-color data and serve as a powerful tool for determining the abundance of ocean biota on a global scale. SeaWiFS also will support operational and commercial activities such as the fishing industry, marine transportation, and offshore oil and gas exploration and development.
- HITC's EOS Data and Information System (EOSDIS) Core System, an electronically accessible global data base. EOSDIS will accumulate, help interpret and distribute EOS information over the initial 10 years and then function for at least 15 years at full capacity to allow accurate modeling of the processes that control our environment.
- SBRC's Moderate Resolution Imaging Spectroradiometer (MODIS), the keystone instrument for the EOS. Scheduled for multiple launches beginning in 1998, MODIS will acquire data that will improve understanding of global dynamics and processes occurring on Earth's surface, in the oceans and in the lower atmosphere.
- Landsat 7, scheduled for launch in 1998, will include SBRC's Enhanced Thematic Mapper Upgrade (ETM+). ETM+ will provide single images covering 31,635 square kilometers of the Earth's surface under Landsat's orbital path, which includes most of the inhabited areas of the world.
- Earth Observing Scanning Photopolarimeter (EOSP), a sensor designed by SBRC, is scheduled to fly aboard EOS AM-2 in 2004.

Hughes' earlier contributions to NASA's Earth resources program include SBRC's Thematic Mappers (TMs), second-generation multispectral scanning sensors flown aboard Landsat 4 and 5, and HSC's Geostationary Operational Environmental Satellites (GOES), which provided timely global weather information, including advance warning of developing storms. Early GOES satellites carried the Visible/Infrared Spin-Scan Radiometer (VISSR), and later GOES satellites carried the VISSR Atmospheric Sounder (VAS), both built by SBRC.

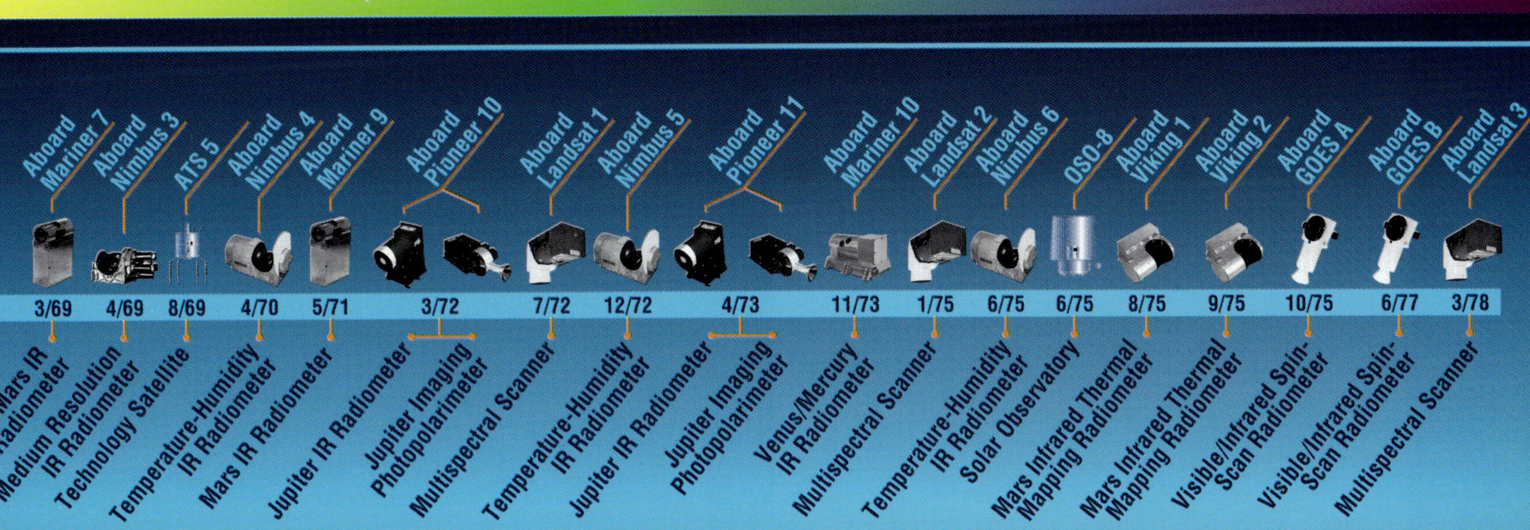

Left to right: A GOES satellite with VAS sensor; the thematic mapper launched aboard a Landsat satellite in 1962; and the Pioneer Venus orbiter, shown here turning its scientific eye toward Halley's Comet in 1986.

Space Exploration

Hughes Space and Communications Company has made major contributions to space exploration with the Pioneer Venus spacecraft and probes, the Magellan radar mapper, the Galileo probe, OSO-8, and Ku-band radar on the space shuttle. SBRC sensors on Hughes and non-Hughes spacecraft also are contributing.

Launched in May and August 1978, the Pioneer Venus orbiter and multiprobe spacecraft arrived at Venus within four days of each other in December. Descending into the Venusian atmosphere, the probes revealed higher concentrations of atmospheric gas than expected and proved water vapor plays an important role in the greenhouse effect of the planet's atmosphere. One of the smaller probes survived impact and continued to send information for 67 minutes. Scheduled to last only one Venusian year (243 Earth days), Pioneer Venus continued to transmit data until October 1992. By the end of its nearly 14-year life, the orbiter had uncovered and transmitted trillions of bits of information, and provided extensive radar mapping of Venus.

Carried into low Earth orbit by the space shuttle in May 1989, Magellan was propelled on its 15-month journey toward Venus by its attached solid-fuel rocket. Between 1990 and 1992, the Hughes-built radar mapper unveiled more than 98 percent of the planet's surface with details as small as 300 feet in diameter. The penetrating radar sensor gave scientists graphic proof of the effects of plate tectonics, volcanism and impact cratering on Venus.

Galileo, launched from the space shuttle in October 1989, will soon complete its journey to Jupiter. In December 1995, the Hughes-built probe will be released for its encounter with the Jovian atmosphere. As the probe begins its fiery descent, the orbiter will receive the probe's science data for retransmission to Earth. The probe will take direct measurements of Jupiter's cloud-shrouded atmosphere and provide data that could help unlock some of the primordial secrets of the solar system.

From a 345-mile high perch above the interference of Earth's atmosphere, the Hughes-built Orbiting Solar Observatory (OSO-8) spacecraft amassed a wealth of data about the sun's radiation. Launched in June 1975, it investigated X-ray and ultraviolet radiation in the turbulent, gaseous band between the sun's surface and the upper reaches of the corona. Designed to

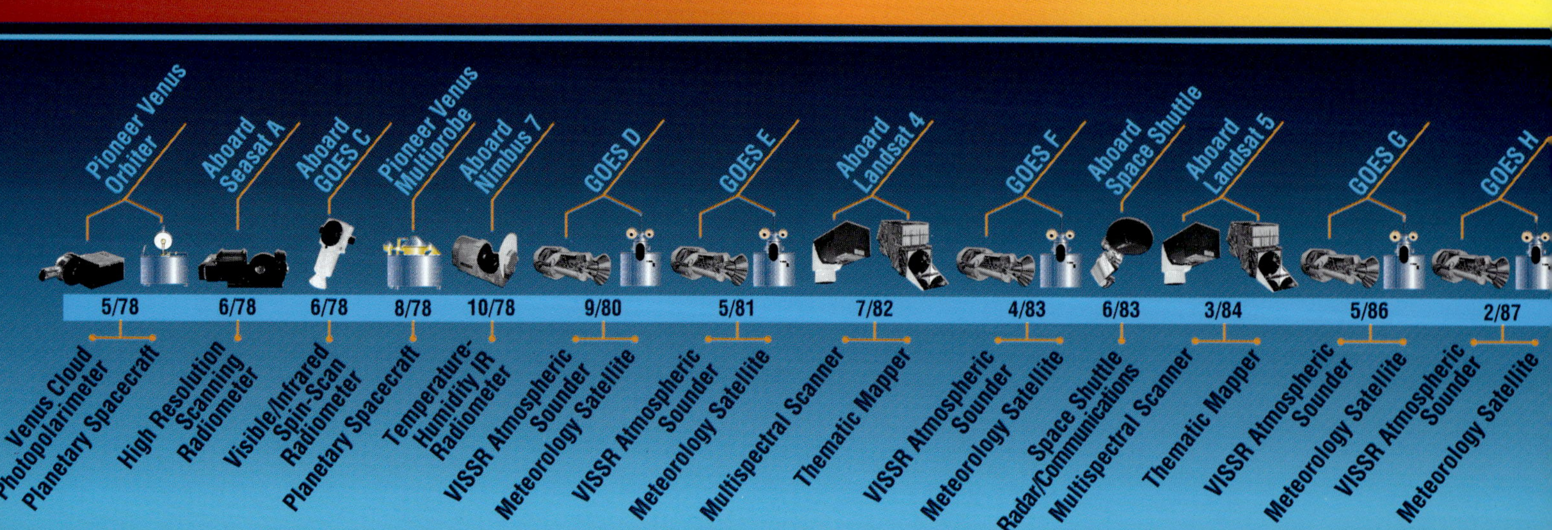

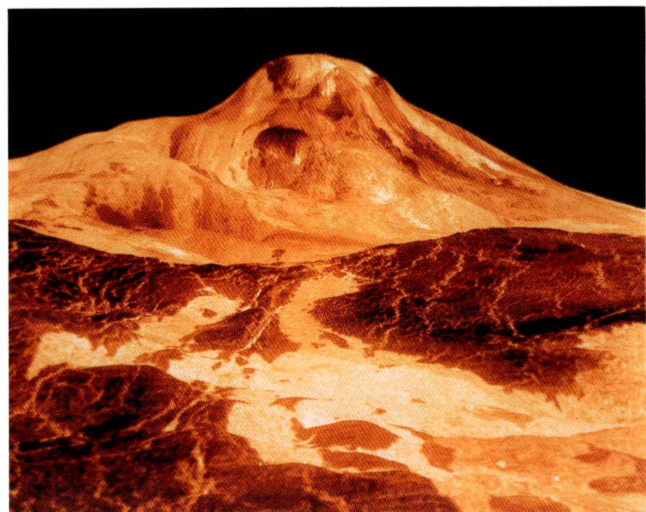

Maat Mons, a 5-mile high volcano on Venus, is shown in this three-dimensional view mapped by Magellan. (JPL photo.)

operate for one year, the spacecraft continued transmitting data until September 1978.

Aboard each space shuttle is the Hughes-built Integrated Radar and Communications System (IRCS), which pinpoints objects in space as far away as 345 miles without help from Earth. This Ku-band radar enables the shuttle crew to locate and rendezvous with objects in low orbit, and has played a critical role in several shuttle missions in which satellites were recovered and either repaired in orbit or returned to Earth. Voice, TV and data links between the shuttle and a ground control center at White Sands, New Mexico, are maintained by the IRCS via a Tracking and Data Relay Satellite (TDRS). In early 1995, Hughes was chosen to provide three next-generation TDRS satellites and upgrades to the White Sands control center.

Space exploration has been further aided by SBRC sensors carried aboard the Vikings (Mars exploration), Mariners (Mars/Venus/Mercury), and earlier Pioneers (Jupiter).

The Thermal Emission Spectrometer (TES), designed by SBRC, is one of a suite of seven instruments on the NASA-JPL Mars Observer Mission. The TES actually comprises three instruments integrated within a single opto-mechanical assembly. The primary TES mission is to study the geology of Mars in detail. The new Mars Global Surveyor Mission also incorporates the TES instrument.

The Galileo Probe, scheduled to descend into Jupiter's fiery atmosphere in December 1995.

Four Decades of Dedication

Now into its fourth decade in space and with more than 700 years of spacecraft operations, Hughes maintains a leadership role in space science. With the dawning of the 21st century, Hughes is poised with NASA to continue a partnership of exploration and discovery to further understand humankind's home in space.

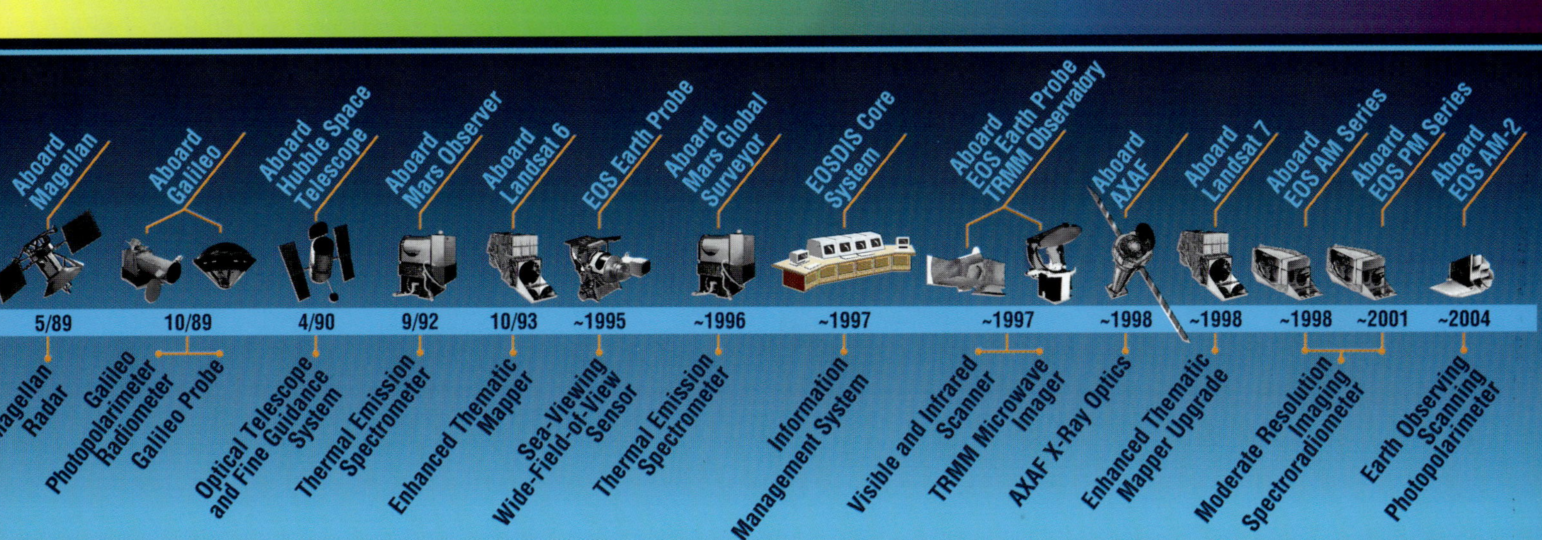

Martin Marietta Corporation

In 1994, scientists at NASA's Jet Propulsion Laboratory in Pasadena, California, plunged the unmanned spacecraft *Magellan* to a fiery death in Venus' dense, acidic atmosphere. Designed and assembled by Martin Marietta, the craft had reached the end of its five-year mission to map the planet, completing what the scientists described as one of the most successful missions in the history of space exploration. *Magellan* had returned more data than all other planetary missions combined.

Released from the cargo bay of space shuttle *Atlantis* on May 4, 1989, *Magellan* was sent on its 795-million mile journey to Venus with the firing of its Inertial Upper Stage rocket booster. Arriving at the evening star in 1990, the $407 million spacecraft transmitted to Earth pictures of an inhospitable planet faced with towering volcanoes, flowing lava and sinuous canyons longer than the Nile River. With *Magellan*'s solar power sails failing after alternately freezing and burning every day for four years, the scientists decided to use its gradual descent to Venus' surface to learn about aerodynamics in the superheated sulfuric mist enveloping the planet. As *Magellan* fired its thrusters three times to lower itself progressively deeper into the planet's atmosphere, the more powerful atmospheric flow from the thicker gas was measured by noting how hard the spacecraft's stabilizing system had to work to keep it from twisting off course. Even in death, *Magellan* continued to provide valuable information.

Magellan, undergoing assembly at Martin Marietta.

Magellan produced clear radar images of Venus's towering volcanoes and deep canyons. Scientists describe the mission as one of the most successful in the history of space exploration.

Four Decades of Space Achievements

Martin Marietta Corporation's involvement in the man and machine quest to conquer space has been ongoing since the dawn of the space age. Four decades of technological developments have made the corporation an industry leader in domestic and international space programs. Today, Martin Marietta Space Group produces the Titan and Atlas family of space launch vehicles, the external tank for the space shuttle and satellites and spacecraft systems for military, civil government and commercial customers. The group provides spacecraft payload integration services and ground-based support for space systems. Headquartered in Bethesda, Maryland, Martin Marietta has operating units and launch facilities in Colorado, Louisiana, New Jersey, Florida, California and Pennsylvania.

Space Group comprises three major operations that offer a diverse combination of space products and services: Astronautics located in Denver, San Diego, Cape Canaveral, Florida and Vandenberg AFB, California; Astro Space, located in East Windsor, New Jersey and Valley Forge, Pennsylvania; and Manned Space Systems, located in New Orleans.

Astronautics

Astronautics designs, develops, tests and manufactures a variety of advanced technology systems for space and defense. Chief products include the Titan and Atlas space launch vehicles, payload integration, the Centaur upper stage and spacecraft for civil and military space programs.

Martin Marietta's Titan family of launch vehicles provides reliable launch options with a heritage of more than 30 years for small, medium and heavy-lift payload requirements. Since the 1960s, Titans have been one of the U.S. Air Force's principal launch vehicles and have launched NASA payloads into Earth orbits and on missions to study the sun and planets.

The first U.S. flight to the moon in more than two decades was launched early in 1994 atop a Titan II rocket. Its payload, the Deep Space Program Science Experiment spacecraft, also called "Clementine," is a joint program of the Ballistic Missile Defense Organization and NASA that mapped the moon's surface. The launch from Vandenberg Air Force Base, California, was the Titan II's fifth space mission since 1986, and nine more Titan II launches are scheduled through the end of the decade. In the 1960s, Titan IIs launched 12 manned Gemini missions for NASA.

Titan IV, which Martin Marietta moved from drawing board to first launch in less than four years, is the largest expendable launch vehicle in the U.S. inventory. First flown in 1989, it is 204 feet long when flown with an 86-foot payload fairing. The system may be flown with a cryogenic Centaur upper stage, an Inertial Upper Stage (IUS), or no upper stage. The first Titan IV with a high-energy Centaur successfully launched the first 10,000-pound Air Force Milstar communications satellite into geosynchronous orbit in February 1994, twice the weight ever delivered to that orbit by any other U.S. expendable launch vehicle.

In October 1997, another Titan IV/Centaur is scheduled to launch the *Cassini* spacecraft on its mission to study Saturn, its moon Titan and the rest of the Saturnian system. A project of NASA, the European Space Agency and the Italian Space Agency, *Cassini* will execute two gravity-assisted flybys of Venus, then one each of Earth and Jupiter before rendezvousing with Saturn in June 2004. Astronautics is performing payload integration, building the propulsion module subsystem, and developing *Cassini's* Descent Imager Spectral Radiometer that will collect data about the atmosphere and surface of Titan, one of at least 18 moons of Saturn.

In 1994, Martin Marietta acquired General Dynamics' Space Systems Division, builders of the Atlas launch vehicles and Centaur upper stages. The Atlas family — Atlas I, II, IIA and IIAS — can lift payloads in the intermediate weight class, generally 5,000 to 8,000 pounds, to geosynchronous transfer orbit. Atlas is a leader in the international and domestic commercial launch business and also counts a number of important U.S. military and government missions in its backlog.

Titan IV, left, will fulfill U.S. launch requirements for heavy payloads beyond the year 2000, until transition to the next generation launch vehicle. Atlas vehicles, including the IIAS, right, launch satellites for the U.S. government and international commercial customers.

Astronautics also developed and builds the Transfer Orbit Stage (TOS) used to boost satellites from low-Earth orbits to higher orbits and planetary trajectories. TOS completed its second mission —first from the space shuttle — in September 1993 when it boosted NASA's Advanced Communications Technology Satellite into orbit from *Discovery*.

Astronautics also designs and builds scientific instruments and experiments. The company's telephone booth-sized Faint Object Spectrograph on the Hubble Space Telescope was used in 1994 to positively identify what astronomers characterize as a "black hole." In late 1995, NASA's *Galileo* spacecraft will drop an expendable probe deep into the turbulent atmosphere of Jupiter where it is expected to transmit data for about 75 minutes before high temperatures and pressures end its mission. Astronautics built three of the probe's six instruments.

Astronautics facilities near Denver house some of the most sophisticated production, space simulation and test facilities in the world. The $676 million physical plant has more than 70 modern laboratories, and computerized engineering design, manufacturing and administrative facilities. Astronautics also operates the U.S. Air Force launch facilities for Titan and Atlas launch vehicle programs at Vandenberg Air Force Base and Cape Canaveral Air Station.

Cassini, to be launced atop a Titan IV/Centaur in October 1997, will arrive at Saturn in June 2004.

Shuttle Orbiter *Discovery* lifts off from Kennedy Space Center in September 1993 with its Martin Marietta-built external fuel tank, Advanced Communications Technology Satellite and Transfer Orbit Stage, which placed the satellite in orbit.

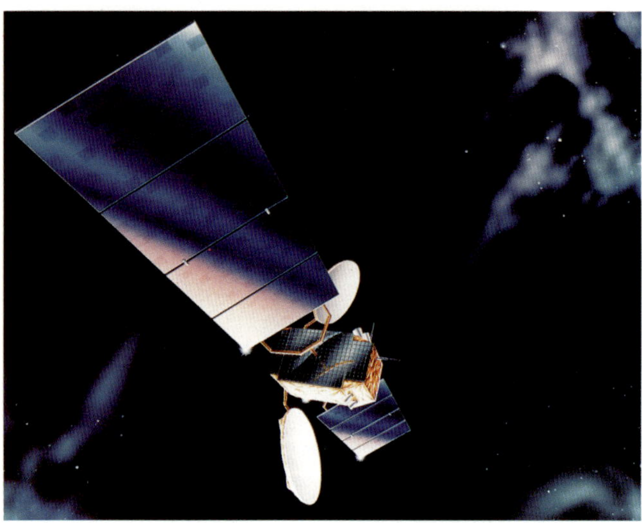

An artist's rendering of the GE-1 spacecraft, built by Astro Space for GE Americom.

Astro Space

Astro Space designs and integrates spacecraft and other space systems for commercial, civil and military customers for missions ranging from scientific to communications to environmental observation to navigation. Key programs include Global Positioning System (GPS), Defense Satellite Communications System (DSCS), Landsat, TIROS and INTELSAT. Astro Space currently has more than 60 spacecraft in various stages of design, development and integration.

Commercial programs in development include AsiaSat-2, EchoStar, GE-1 and -2, Inmarsat-3, INTELSAT VIII and VIII-A, Koreasat, and Telstar 4. Telstar 401, the first of three advanced communications satellites developed for AT&T, was launched in December 1993. It is the first Series 7000 spacecraft, the newest and largest of Martin Marietta's family of communications satellites, which feature arcjet thrusters that consume less fuel for station-keeping, thereby extending spacecraft life, and variable-output, solid-state power amplifiers, which offer increased mission flexibility. INTELSAT, the International Telecommunications Satellite consortium, has ordered six Series 7000 satellites. Launched in 1994 was the BS-3N television direct broadcast satellite for Japan. The first of five Inmarsat-3 spacecraft for the International Maritime Satellite Consortium is scheduled for launch in late 1995.

A Defense Meteorological Satellite nears completion at Astro Space.

Current government programs include Advanced Communications Technology Satellite (ACTS), Global Geospace Science (GGS), NOAA/NASA polar-orbiting environmental satellites (TIROS), Space Video, *Cassini* RTGs, and Mars Global Surveyor Subsystems. Wind, the first GGS satellite for NASA, was launched in 1994.

Current defense programs include Defense Satellite Communications System (DSCS III) satellites and ground system, Defense Meteorological Satellite Program (DMSP), Global Positioning System (GPS IIR) satellites, hypersonic vehicles and Milstar subsystems. Continuing a record of communications support to the U.S. Armed Forces, uninterrupted since 1982, two DSCS III spacecraft were launched and successfully deployed in 1993.

Astro Space's long involvement in building environmental satellites continues with EOSAM-1, the first satellite in NASA's Earth Observing System, due to be launched in 1998. Since 1960, more than 70 Astro Space satellites have provided a continuous study of the Earth's environment. Other remote sensing programs include continuing development of Landsat 7 for Earth observation. Launched in 1994 were DMSP 30 and NOAA-14, the latest in a series of more than 50 civil and military weather satellites that will continue into the next century.

Astro Space communications satellites have accumulated more than 200 operating years, and by the end of 1996 will be serving a minimum of 12 systems. Development continues on the A2100 spacecraft, which will serve as a basic platform for the next generation of communications satellites. The A2100 will offer increased capacity at lower weight, fewer parts and reduced time between order and delivery, which means lower cost without sacrificing orbit life. GE-1, the first A2100, is due to be launched in 1996 for GE Americom.

Manned Space Systems

As prime contractor on the space shuttle External Tank project at NASA's Michoud Assembly Facility in Louisiana, Manned Space Systems performs design, development, test and engineering and production of the space shuttle's 154-foot-long, 535,000-gallon external tank. Under existing contracts, Martin Marietta will continue to produce the tanks into the year 2000.

The external tank is the largest element of the shuttle and has a dual role during the first eight and one-half minutes of each launch. It feeds cryogenic liquid propellants to the three main engines and serves as the primary structural backbone of the vehicle, absorbing four million pounds of thrust. After main engine cut-off, the tank separates from the shuttle and falls back into the Earth's atmosphere, disintegrating harmlessly over the ocean.

Major innovations under study by Manned Space Systems for NASA include a new super lightweight external tank, and through a Technology Reinvestment Project, the development of hybrid propulsion technology that will use both cast solid fuel and cryogenic liquid oxidizer in a single motor.

Manned Space Systems' unblemished record of external tank successes has been recognized by the receipt of NASA's first Excellence Award for Quality and Productivity, the U.S. Senate Productivity Award and the NASA Flight Safety Award.

Other Space Programs

In 1993, Martin Marietta won a nine-year follow-on contract to support NASA's Life Sciences Directorate at the Johnson Space Center. The scope of the program was increased to include support for mission integration for nine dockings between the space shuttle and the Russian *Mir* spacecraft. As a part of this contract, Martin Marietta Services Group provided substantial support to the October 18, 1993, space shuttle mission dedicated to understanding the effects of weightlessness on human health and productivity. Services Group also designed, built and tested major systems for the Spartan 201 science satellite launched in April 1993.

As a subcontractor to McDonnell Douglas, Martin Marietta Communications Systems, part of the Electronics Group, is designing three communications subsystems for the Space Station. These systems will control all of the station's radio communications and external video.

Martin Marietta Management and Data Systems, part of the Information Group, has a long-term contract to redesign a satellite support system. Management and Data Systems has more than 20 years experience with the system and is developing a state-of-the-art distributed computer system to modernize the entire network.

Tracking and Data Relay Satellite System ground antennae.

Space shuttle's external tanks are built by Manned Space Systems at NASA's Michoud Assembly Facility in Louisiana.

With significant contracts from the International Maritime Satellite Organization, INTELSAT and AT&T, Martin Marietta has established itself as a commercial systems contractor for communications satellites. The corporation is also building a turnkey satellite command and control system in support of Orion Satellite Corporation's first communications satellite, launched in 1994.

Technology for the Future

Martin Marietta's Advanced Development Operations and Laboratories work with operating units to formulate shared visions of new products that anticipate future requirements in space, defense, information systems and commercial markets. From research in electronics, materials and information processing, Martin Marietta scientists and engineers focus on developing products and technologies that meet these needs.

The research and development operations from Martin Marietta and former GE Aerospace are consolidated under a single technology organization with operating elements in California, Maryland, New Jersey and New York.

Martin Marietta Laboratories in Baltimore is leading a multi-partner team comprising aerospace firms, suppliers, universities and government laboratories that received one of the first contracts to be awarded under the U.S. Government's dual-use technology initiative. The program will develop low-cost, broadband, active anti-vibration techniques, utilizing company expertise in ceramics and shape memory alloys.

In 1993 and 1994 respectively, Martin Marietta Corporation acquired General Electric's aerospace businesses and General Dynamics' space systems unit. Resultant consolidations and cost-reduction initiatives are benefiting the national space and defense programs as Martin Marietta continues development of next-generation satellite platforms, spacecraft systems and launch vehicles.

In 1995, Martin Marietta Corporation will merge with Lockheed Corporation, which had purchased General Dynamics' F-16 fighter aircraft business. The new corporation will be a highly diversified, advanced technology company with core businesses in defense, space, energy, information, commercial, civil government, and international markets. The new company, Lockheed Martin Corporation, will have annual sales of nearly $23 billion and employ approximately 170,000 people.

Motorola

Manned or unmanned, from near Earth orbit to deep space, spacecraft maintain a vital link to Mother Earth via Motorola equipment.

Motorola equipment has contributed to the success of every manned U.S. space venture and a majority of the unmanned U.S. space programs since the earliest days of America's space program. In the early 1960s, the company provided the command receiver for the Mercury capsule that reliably served all Mercury astronauts. For the Gemini series, Motorola developed and produced the spacecraft's digital command system, which received Earth-based signals, decoded and passed them to onboard subsystems, then obtained the subsystem "answers" and relayed them back to Earth.

Before the Apollo astronauts journeyed to the moon, Motorola transponders performed command, tracking and telemetry functions on the unmanned Ranger spacecraft that made extensive close-up studies of the moon's surface and paved the way for Apollo. On the way to the moon, the company's S-band transponder provided the only means of communication between the astronauts and ground control once the spacecraft had traveled more than 45,000 kilometers from Earth. From the moon's surface, the sounds of the astronaut's voices, biomedical data, and television signals were transmitted back to Earth via a Motorola transceiver on the lunar module and an S-band transponder on the command module. As Apollo astronauts drove the lunar rover on the moon's surface, they received communications from Earth on the Motorola FM receiver. The employment of that sophisticated equipment on the moon created a milestone in the history of Motorola, the company that developed and perfected the first automobile radio on Earth.

Two Motorola S-band transponders supported the Apollo-Soyuz docking in space — one was on the Apollo for communication with ground control, the second permitted Americans to view the actual docking. After the Apollo series of missions ended, Motorola supplied space communications equipment for the Skylab orbiting laboratories.

Motorola equipment on Mariner spacecraft transmitted pictures of Mars to Earth from a distance of 200 million miles. From the Viking spacecraft — one that landed on Mars and another that orbited the planet, pictures were transmitted from the Martian surface to the orbiter and relayed back to Earth via Motorola equipment. The two Voyager spacecraft have transmitted pictures of Jupiter, Saturn, Uranus and Neptune back to Earth, and Voyager II, traveling more than 1.8 billion miles into space, still communicates with Earth via its Motorola communications system. The system, which is comprised of a transponder,

The S-Band Receiver produced by Motorola for the Lunar Roving Vehicle was the first car radio on the moon and the only voice link the astronauts in the Rover had with Earth.

Motorola equipment on Mariner II, IV and X transmitted pictures of Mars, Mercury and Venus to Earth.

control unit, ultra-stable oscillator, solid-state microwave switches and traveling wave tube amplifiers, provides the two-way communications link between Voyager spacecraft and Earth.

The Magellan Venus probe, launched from space shuttle in 1989, completed its comprehensive observation of Venus' surface and gravitational features in 1994. It carried the Motorola Radio Frequency Subsystem (RFS), which received all command and control information and converted it to digital data for the spacecraft's flight computer. The RFS also transmitted all engineering and scientific data from the spacecraft back to tracking stations on Earth. Magellan is considered the most successful interplanetary mission to date.

Motorola has manufactured highly reliable Telemetry, Tracking and Command (TT&C) transponders for spaceflight missions for more than 35 years. These transponders provide the primary communication link between spacecraft and Earth. Motorola's Spaceflight Tracking and Data Network (STDN) transponder provides S-Band communications for NASA and the European Space Agency (ESA) missions.

Early participation in evaluating user equipment complexity versus various spread spectrum formats and techniques led to Motorola's development of an S-band TDRSS (Telemetry, Data, Relay Satellite System) User Transponder, later known as the NASA Standard TDRSS Transponder. Motorola also developed and delivered the Master Frequency Generator and the 32-channel Multiple Access Receiver for the Tracking Data Relay Satellites (TDRS) developed by TRW. Today, Motorola's third generation of TDRSS user transponders use digital signal processing technology that improves reliability, radiation hardness and receiver power dissipation. Since the beginning of the space shuttle flights in 1981, TDRSS has enabled astronauts to communicate with Earth from nearly any orbit.

The Cassini X-Band Transponder, designed by Motorola and the Jet Propulsion Laboratory (JPL), meets the most demanding requirements for missions anywhere in the solar system and beyond. Advanced frequency synthesis technology employing dielectric-resonator oscillators, temperature-compensated crystal oscillators, and high-order, phase-locked multipliers has resulted in a new standard for receiver stability, exciter spurious, and exciter phase-noise. State-of-the-art components optimize the performance of critical circuits such as the receiver's X-Band low-noise amplifier and to minimize size, mass and power.

Motorola transponders have been used on various spacecraft missions including the Hubble Space Telescope, the NASA Explorer Platform Program (NEPP), NASA's Ocean Topography Experiment (TOPEX), Gamma Ray Observatory, and Cosmic Background Explorer.

Motorola's Monarch™ Global Positioning System (GPS) spaceborne user receiver enables continuous autonomous spacecraft navigation with position, velocity, and time information of unprecedented accuracy. The second-generation Monarch has all the performance characteristics of the original, but is much smaller and can be implemented for multiple antennas simultaneously. Viceroy™, a smaller version of the original Monarch, operates at reduced power and with dual antennas that improve visibility, acquisition time and performance. The GPS spaceborne receiver first flew aboard NEPP in 1991 and has application in virtually every spacecraft program of the future.

Motorola's Crosslink Transponder and Data Unit (CTDU) is a time division multiplexed-frequency hopped spread spectrum communications transponder that provides a communication link between satellites in the Global Positioning System. The CTDU transfer data in three different modes in addition to measuring

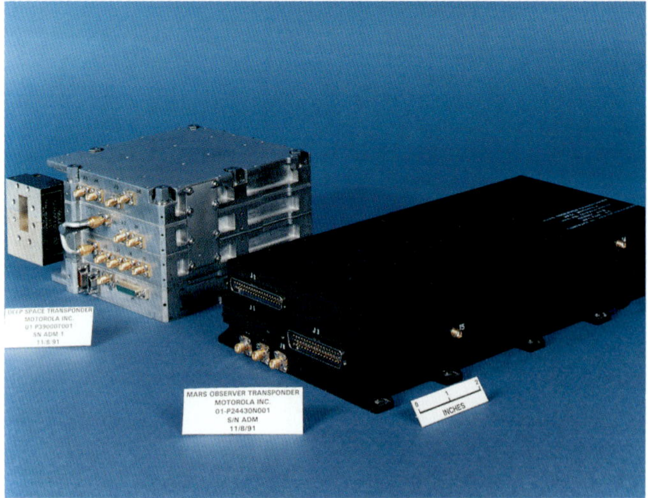

From Near Earth Orbit to Deep Space, spacecraft maintain a vital link to Earth via Motorola transponders.

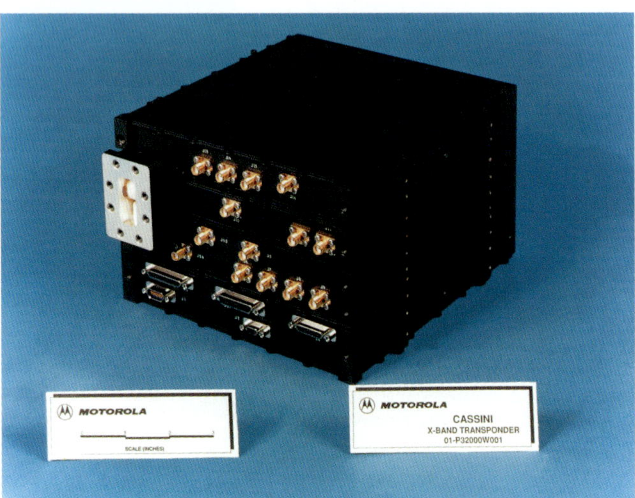

the range between satellites. State-of-the-art digital signal processing techniques are used to provide optimum performance while minimizing size, weight and power.

Motorola's Baseband Processor (BBP) is the core of NASA's next-generation communication satellite payload. Capable of handling the equivalent of 66,000 telephone calls simultaneously, Motorola's BBP was adopted by NASA as the heart of the Advanced Communications Technology Satellite (ACTS). ACTS is an experimental tool used by U.S. corporations, universities and government agencies as a vehicle for evaluating technologies that are key to the future growth of satellite communications.

For space station, Motorola's baseband signal processors will provide video, audio and high-rate data processing in conjunction with the Motorola Ku-band TDRSS transceiver modem and standard TDRSS transponder for the space-to-ground, space-to-space and assembly/contingency subsystems. Motorola's space data interface module links space experiments to the EOS and SSF platforms. Other Motorola equipment with space application are the company's modular DC-DC power converter subsystems, which provide maximum versatility in power level management, heat distribution and reliability, and the Multi-Mode PCM Demodulator/Equalizer, which provides a high degree of flexibility to meet PCM demodulation needs.

Motorola's aerospace operations fall under the Government and Space Technology Group (GSTG), which has repeatedly produced technological breakthroughs in a wide variety of electronic components, products and systems since the 1950s. GSTG today has world-class competencies in communications systems, system security engineering, tactical electronic systems, software systems, and decision systems.

In addition to the many products and systems developed for the space program, GSTG has solved complex communication challenges for commercial, civil, governmental and international customers. Thousands of clear and secure voice and data communication products and systems have been developed, manufactured and fielded for land-based, mobile, shipboard, airborne and space-based installations.

Motorola GSTG's extensive experience in defense and space electronics, along with corporate leadership in commercial wireless communication, allows smooth and rapid transition from military technology to commercial applications (spin offs) and commercial solutions to government needs (spin ons). As a spin off example, the technology and knowledge gained during the design of hardware and software for the Department of Defense's Global Positioning System satellites was instrumental in the development of several commercial GPS-based location and tracking systems.

Expertise in systems management, data transmission and satellite communications enabled GSTG to develop the IRIDIUM™/SM System, a seamless, wireless, digital network based on a constellation of 66 satellites in low-Earth-orbit, which permits any type of telephone transmission — voice, data, facsimile or paging — to reach its destination anywhere on Earth, at any time.

Today, GSTG continues to research, develop and produce high-technology terrestrial and space-based communications equipment and systems, sensors and specialized electronics for customers worldwide.

The Motorola-developed IRIDIUM system is a seamless, wireless, digital network based on a constellation of 66 satellites in low-Earth-orbit which permits any type of telephone transmission to reach its destination anywhere on Earth at any time.

Motorola History

Motorola was founded by Paul V. Galvin in 1928 as the Galvin Manufacturing Corporation in Chicago, Illinois. Its first product was a "battery eliminator," which allowed consumers to operate radios directly from household current instead of the batteries supplied with early models. In the 1930s, the company successfully commercialized car radios under the brand name "Motorola," a new word suggesting sound in motion. The name of the company was changed to Motorola, Inc. in 1947.

In the 1940s, the company entered government work and opened a research laboratory in Phoenix, Arizona to explore solid-state electronics. By the time of Paul Galvin's death in 1959, Motorola was a leader in military, space and commercial communications, had built its first semiconductor facility and was a growing force in consumer electronics.

Under the leadership of Robert W. Galvin (Paul Galvin's son), Motorola expanded into international markets in the 1960s and began shifting its focus away from consumer electronics. The color television business was sold in the mid-1970s, allowing Motorola to concentrate on high-technology markets in commercial, industrial, and government fields.

Today, Motorola is one of the world's leading providers of wireless communications, semiconductors and advanced electronic systems and services. Major equipment businesses include cellular telephone, two-way radio, paging and data communications, personal communications, automotive, defense and space electronics and computers. Communication devices, computers and millions of consumer products are powered by Motorola semiconductors. Annual sales are approximately $20 billion.

Motorola's operations are highly decentralized into four sectors and two groups: Semiconductor Products Sector; General Systems Sector; Land Mobile Products Sector; Messaging, Information and Media Sector; Government and Space Technology Group, and Automotive, Energy and Controls Group. The company maintains sales, service and manufacturing facilities throughout the world, conducts business on six continents and employs approximately 120,000 people worldwide. Total customer satisfaction is the company's fundamental objective.

Network Systems Corporation
High-Performance Networking Solutions

Whether managing data from satellites deep in space or for organizations with locations around the world, Network Systems Corporation is providing the products and services that make up the backbone of high-performance network solutions. Founded in 1974, Network Systems Corporation pioneered the computer networking industry and today remains the market leader in channel networking.

Network Systems is a major supplier of internetworking routers and hubs, and plays a leading role in network design and implementation — from internetworking PC LANs to gigabit speed networks that can cross continents. For 20 years, the company has worked side by side with customers, listening to their IS concerns and responding with leading edge technology.

Solutions in Network Systems' portfolio include client/server computing, disaster recovery, data center consolidation, virtual networking, network-attached storage, networking branch offices, remote device support, telecom services, and world-wide access to the full range of resources necessary to manage networks.

Space Network Support

Since 1987, scientific data from satellites in deep space and from Spacelab experiments aboard the space shuttle has been processed by the MODNET HYPERchannel networking system established by Network Systems Corporation at Goddard Space Flight Center in Greenbelt, Maryland.

The Mission Operations and Data Systems (Code 500) Directorate at Goddard processes data from unmanned satellites and controls the attitudes and orbits of those satellites. The consolidated MODNET, which replaced three separate HYPERchannel computer networks, has the capability of transferring large amounts of data from computer to computer with no delays, no bottlenecks, no compromises in the quality of the data — and with speed.

MODNET initially connected 34 large computers of various types (IBM, NAS, DEC, DEC VAX, Perkin-Elmer, Gould SEL, and UNISYS) located in three different buildings on the Goddard campus, and its modular design allowed room for others to be added as needed. Network Systems' HYPERchannel network processors, high-performance computers that

Scientific data from deep space satellites and from Spacelab experiments aboard the space shuttle has been processed since 1987 by the MODNET HYPERchannel networking system established by Network Systems Corporation at Goddard Space Flight Center in Greenbelt, Maryland.

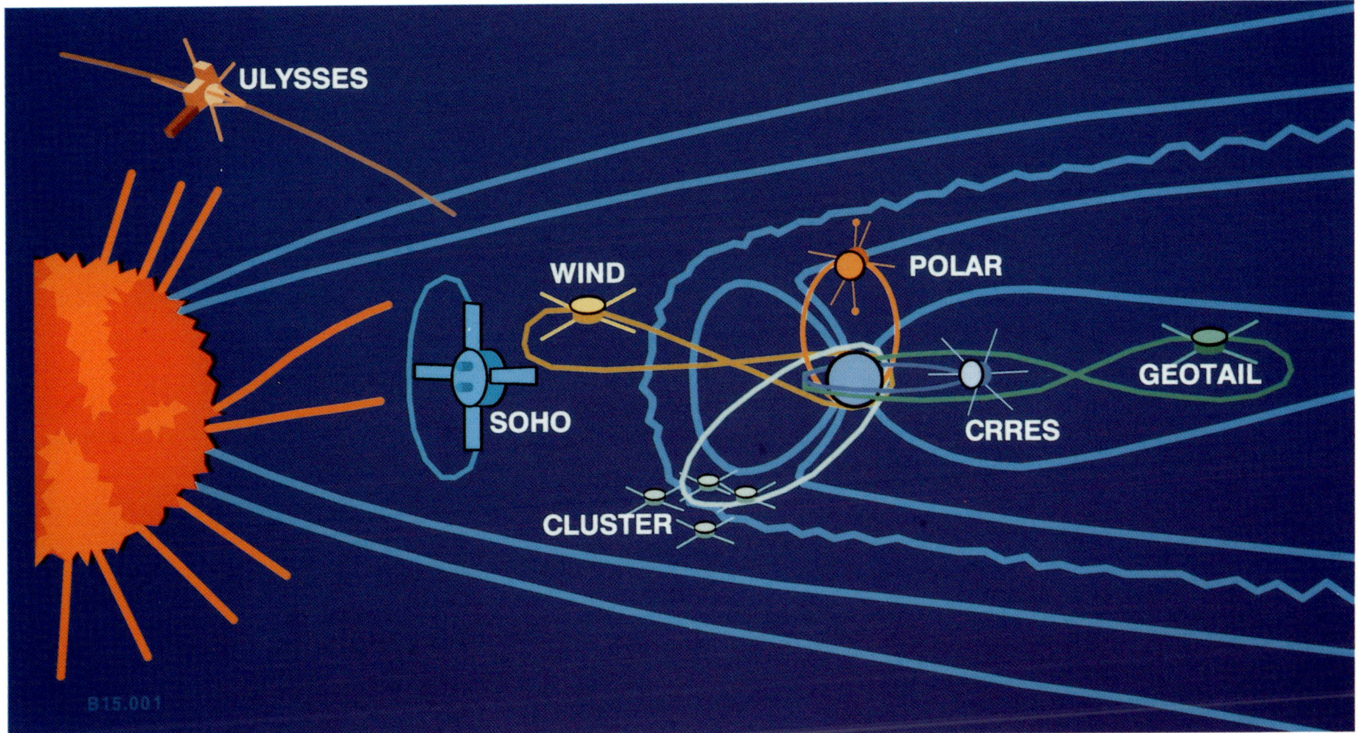

This mission configuration of the International Solar-Terrestrial Physics (ISTP) Program is designed to draw on the resources of a worldwide scientific community to make a concentrated and coordinated study of the interactions in the Sun-Earth system, and to extrapolate this knowledge to other planets and to the universe beyond. Data collection for this ambitious program is the responsibility of Goddard Space Flight Center's Mission Operations and Data Systems Directorate and its MODNET HYPERchannel network.

move data between computers, peripheral devices, long-distance communications links, or between entire networks, was the selected hardware. The company's NETEX (NETwork EXecutive) communications software provided access from one computer to the other, regardless of the type of computer or operating system.

Fiber optic cable connected two of the buildings in the network, and coaxial cables for other trunk connections to the MODNET system. For these connections, Network Systems supplied cable connectors, terminator blocks and fiber optic repeater devices. Two of the four trunks connected to each network provide local or intradivisional access, and two provide for interdivisional "highway" trunks that handle only operational data — orbit, attitude and telemetry data gathered by the satellites.

One segment of the MODNET network, the MODLAN subnetwork, is accessible only through an electronic bridge, allowing electronic isolation. Another subnetwork, INFOLAN, provides both optical mass storage for network data and a gateway to the Space Physics Analysis Network (SPAN) for the transfer of science-related data to research centers and other NASA sites. To prevent infection of the entire MODNET network by possible "computer viruses," three isolated computers accept data from outside networks and have to be manually unlocked to allow information to flow into MODNET.

Because NETEX and the HYPERchannel network processors can have their operational parameters modified at any time to fit the changing needs of the network, the host computers can be switched from one mission or task to another as required by the Code 500 Directorate.

Mission Success

Information processed by MODNET has continually expanded and changed human concepts about Earth and the universe.

The International Solar Terrestrial Physics (ISTP) program, seven satellites from the United States and Japan, are conducting a comprehensive program of solar-terrestrial research to measure, model and assess quantitatively the process in the Sun-Earth interaction chain. These satellites are gathering data on the deposition of energy into the atmosphere; global plasma storage, flow, and transformation; solar wind magnetosphere interactions; solar wind origins and three-dimensional features; coronal dynamics; solar seismology; and basic plasma states and characteristics. ISTP sends 4.2 gigabytes of data each day to be processed on MODNET.

Positioned above Earth's atmosphere, COBE (Cosmic Background Explorer), sent into orbit in 1989, has a prime viewing position for collecting data and sending it back to Earth. Its data, sent via the orbiting Tracking and Data Relay Satellite System (TDRSS) to the NASA ground receiving station at Wallops Flight Facility in Virginia, is being analyzed and converted into maps of the sky. COBE's record of

Hubble Space Telescope is providing astronomy's newest and clearest window on the universe. Data from Hubble is being processed by the Network Systems Corporation's MODNET.

Data from deep space satellites relayed via the orbiting Tracking and Data Relay Satellite System (TDRSS) to the NASA ground receiving station at Wallops Flight Facility in Virginia is processed by MODNET.

the radiation echoes of the universe is challenging scientists' understanding of the Big Bang theory of the exploding universe.

The repaired Hubble Space Telescope, also positioned above Earth's atmosphere, is providing astronomy's newest and clearest window on the universe. Data from Hubble, as well as from LAGEOS (the Laser Geodynamics Satellites, a cooperative study with Italy of crustal dynamics), and from the Upper Atmosphere Research Satellite (UARS), studying the mechanisms controlling the upper atmosphere, is being processed by the Network Systems Corporation's MODNET.

Code 500 supports the scientific Spacelab experiments aboard the space shuttles. Through SIPS (Spacelab Input Processing System), scientific data from the shuttles is entered into the MODNET network. Through SOPS (Spacelab Output Processing System), data is prepared and sent to users at principal research laboratories throughout the United States. Other space shuttle missions, such as the servicing of satellites while in orbit (for example, Solar Maximum Mission and the repair of Hubble) are supported by Code 500 and MODNET.

Astronaut George D. Nelson moves in to service the Solar Maximum Mission satellite. Space shuttle missions to service satellites in orbit are supported by Code 500 and MODNET.

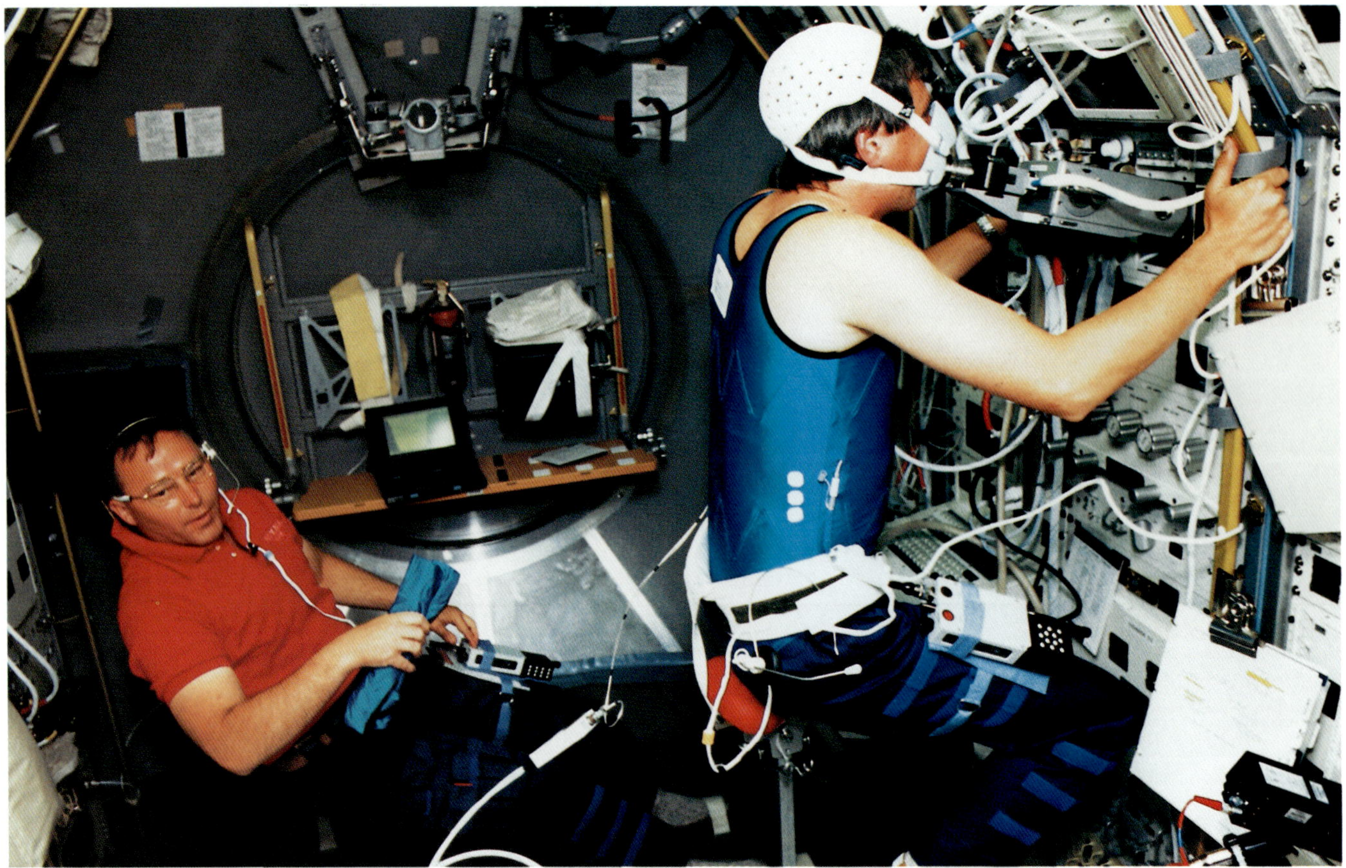

Ulrich Walter (right) serves as a test subject to investigate human physiology under microgravity conditions at the Anthrorack in the science module onboard the Earth-orbiting space shuttle Columbia during a 10-day Spacelab D-2 mission in April-May 1993. Astronaut Jerry L. Ross, payload commander, assists in the test run. Data from Spacelab science missions is processed by the MODNET HYPERchannel networking system at Goddard Space Flight Center.

HYPERchannel Speeds at Other NASA Sites

For 20 years, Network Systems Corporation's hardware and software has been used to build high-performance computer networks at a number of NASA sites.

At Marshall Space Flight Center's Michoud Assembly Facility in New Orleans where the external tanks used in space shuttle launches are built, HYPERchannel-DX hardware serves as the backbone of a 100-Megabit fiber optic network that uses Ethernet and connects ComputerVision, Data General, Wang, IBM, DEC and UNISYS computers.

At NASA Langley in Virginia, Network Systems hardware and software have supported the flight simulation cockpit since 1979, carrying data in real time from the cockpit to a DEC minicomputer to a Control Data CYBER mainframe, back to the DEC, and then to the cockpit. Langley also uses Network Systems' equipment in a high-performance data highway that connects a Cray supercomputer to IBM, Convex and Sun systems.

Six major laboratories at Marshall Space Flight Center in Huntsville are connected by Network Systems technology to a high-powered central computing system that includes a Cray and two IBM computers. Network Systems also provides the high-speed backbone for MARS, the Marshall Archival Retrieval System.

A New Generation of Networking Products

Committed to maintaining its position as the leader in providing market-driven solutions for high performance, cost-effective, enterprise-wide information networks, Network Systems at the end of 1994 had in place a complete new generation of networking products. These products adhere to a core set of industry standards, including ATM, Ethernet, FDDI, Frame Relay, HIPPI, NetView, OSI, SMDS, SNA, SNMP, TCP/IP, and Token Ring.

Products include:

- Unlimited-distance channel extension and CPU networking devices
- Interconnect controllers that connect LANS to mainframes
- RISC-based, multi-protocol bridge-routers for LAN, WAN, and MAN connections
- Intelligent port-switching hubs
- High-speed switches compatible with the HIPPI standard
- Software for centralized network backup, remote printing, file transfer, TCP offload, security and virtual networking

Network Systems provides technology leadership as a founding member of the National Storage Laboratory, a consortium implementing the IEEE Mass Storage Reference Model, and as a member of ATM

Forum, ATM Standards Committee, American Standards Committee for OSI Network and Transport, HIPPI and Fibre Channel Standards Committee, Internet Engineering Task Force, NIIT (National Information Infrastructure Testbed-founding member), Open Data Center Alliance, OSPF interoperability Working Group, PPP (Point-to-Point Protocol) Consortium, SMDS Interest Group, and SONET Standards Committee.

Strategic partnerships with leading manufacturers and industry groups enables Network Systems to share technology, resources, products and understanding. Alliances have been fostered with Amdahl, Bell Atlantic, Bull, Centel, Comidisco, Cray Research, Digital Equipment Corporation, Digital Link, Hewlett-Packard, IBM, ICL, Maximum Strategy, NAI Technologies, Northern Telecom, ODS, Storage Technology, SunGard, Tandem, Unisys, US West, and WilTel.

Acquired in 1993, Bus-Tech, Inc., a subsidiary, provides product solutions that enable customers to connect their LANs to mainframes, and Bytex, a strategic business unit, adds intelligent port-switching hubs to implement virtual networking at the network level and network management software that can be adopted across Network Systems' entire product family. Bus-Tech and Bytex are major resources for future product development.

Serving Customers Worldwide

Network Systems experience and expertise is being used by more than 5,000 customers worldwide. In addition to NASA, customers who apply Network Systems' solutions include telecom carriers, other government agencies, large manufacturers, corporations with widely distributed operations such as retailing, banking and insurance, and research institutions and national labs.

Headquartered in Minneapolis, Minnesota, Network Systems has 1200 employees and more than 60 sales and service locations worldwide. The company's products are sold and serviced through direct sales in North America and Europe, and distributors in Europe and Asia/Pacific.

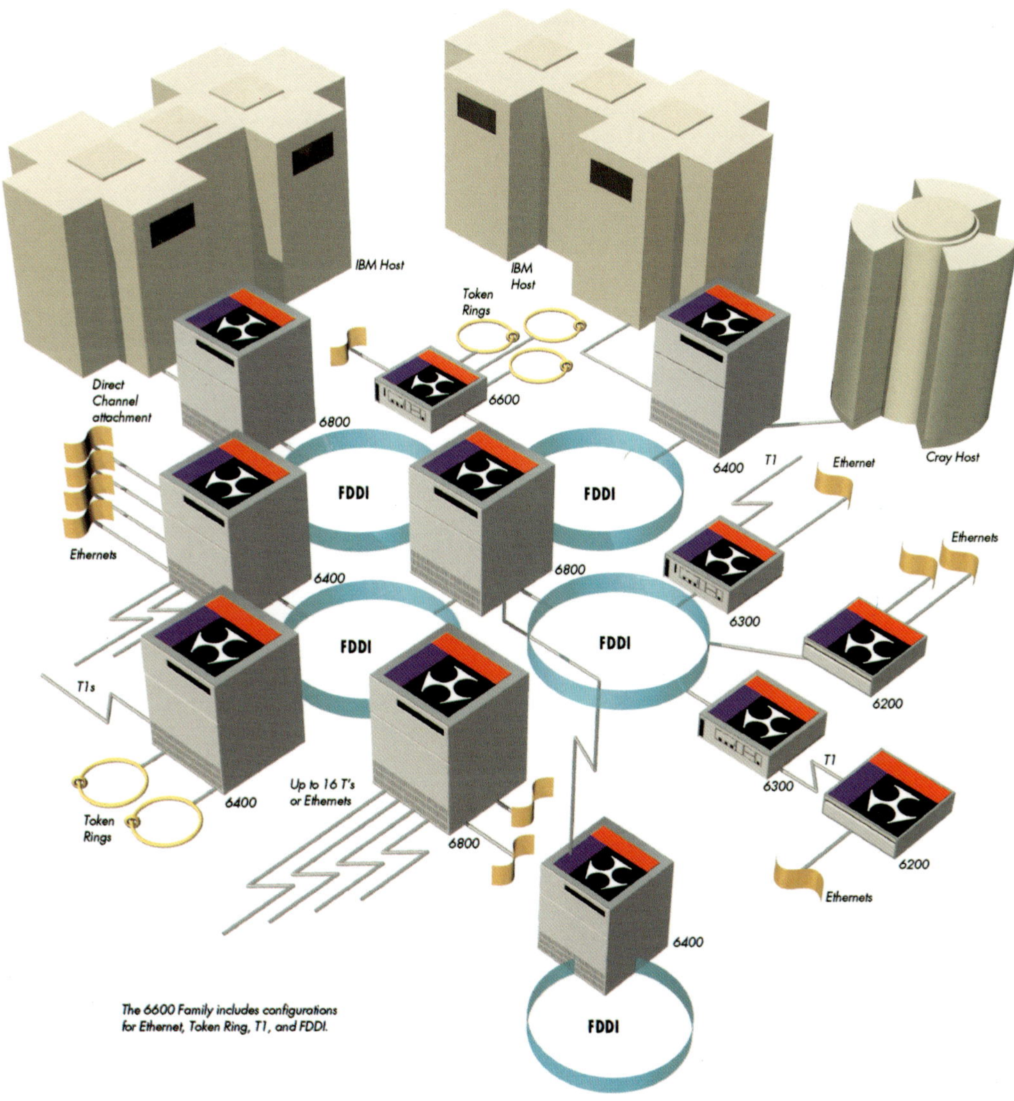

Network Systems Corporation introduced a complete new generation of networking products in 1994. The 6600 family shown includes configurations for Ethernet, Token Ring, TI and FDDI.

DATATAPE Incorporated

DATATAPE Incorporated has played a key role in the nation's space program since 1959 when "Project Mercury" was commissioned with sending American astronauts into space. So it was that on May 5, 1961, as Lt. Comdr. Alan B. Shepard, Jr. was launched into space aboard Mercury shouting "Man, what a ride," a DATATAPE recorder weighing less than nine pounds was measuring the responses of both the astronaut and his spacecraft while transmitting them to 18 tracking stations around the world. Twenty years later, on April 12, 1978, America's Space Shuttle made its first successful flight into orbit and DATATAPE again was there with equipment to record flight systems data as well as the astronauts' respiration, EKG and psychomotor reactions.

In the two decades between the first manned space flight and the Space Shuttle, DATATAPE remained heavily involved in many aspects of the space program. When Neil Armstrong became the first man to set foot on the moon in 1969, a significant amount of his mission's data evaluation and communication relied on DATATAPE equipment at Mission Control in Houston. In fact, DATATAPE products have supported every phase of America's manned space program from Gemini and Apollo to Skylab and Space Shuttle.

As the space programs have advanced so have DATATAPE instruments advanced, keeping time with the rapidly changing requirements of recording more data at higher rates with greater reliability. Recorders were developed to perform functions such as recording critical flight test data on the Saturn rocket during Apollo, and now on the shuttle's solid-rocket boosters (SRBs). The specific products that have served and continue to serve America's space program include the DDR-100 and MARS 2000 and 1400 Series. Depending on each shuttle's specific mission, any of these DATATAPE recorder/reproducers might be at work. For example, DATATAPE's DDR-100 was the recorder of choice in the Modular Optoelectronic Multispectral/Stereo Scanner (MOMS-02) because of its high recording/data rate, low weight and low power consumption, characteristics which are essential requirements in space research. MOMS-02 was carried into orbit by Space Shuttle *Columbia* in 1993 and successfully gathered data for use in developing the next generation maps of Earth's topology, vegetation, rock formations and soils. MOMS-02 successfully implemented new technologies that combined within a single system the capabilities for generating both high-resolution panchromatic images (which give highly precise, three-dimensional, geometric information) and multispectral images for thematic mapping. These maps will not only benefit the scientific community, but also aid the layman's understanding of the environment.

DATATAPE's dedication to quality products and service to the space program were highlighted March 17, 1994, when NASA astronaut Catherine Coleman presented three DATATAPE employees with prestigious Silver Snoopy awards for overall excellence in job performance contributing to the success of the nation's space program. These awards are given to individuals who demonstrate highly creative solutions, exceptional performance and who continue to provide customer assistance above and beyond the call of duty.

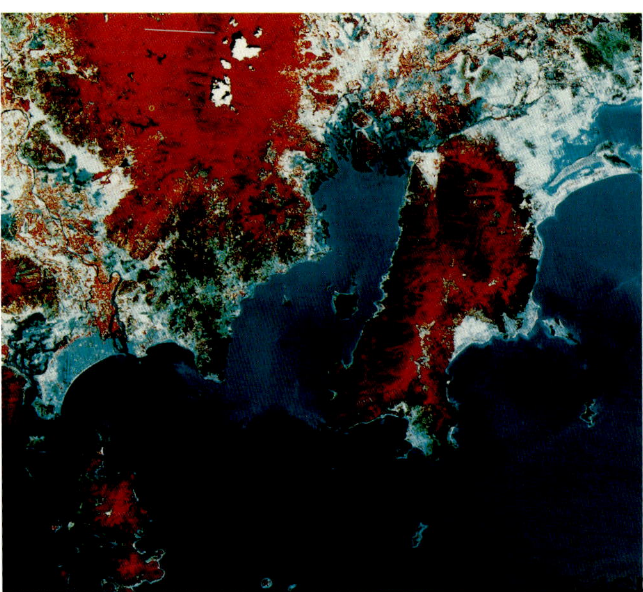

The data gathered with the Modular Optoelectronic Multispectral/Stereo Scanner (MOMS-02) took more than six months to translate and will be used for making the next generation land maps of the Earth's topology. Photos courtesy of Deutsche Aerospace, AG (DASA).

Three DATATAPE employees were honored by NASA and recognized for their outstanding contributions to the nation's space program. Shown left to right are George Walters, Catherine Coleman (NASA astronaut, presenting the awards), Gary Carr and Arthur Snook.

DATATAPE Looks to the Future

DATATAPE continues in its quest for delivering advanced technology recorder/reproducers and has taken significant steps over the past few years in a commitment to being a highly dynamic, market-driven business. The most significant change occurred in 1994 when DATATAPE was spun out of Kodak, becoming an independent small business. New Chairman and Chief Executive Officer Dom Saccacio is targeting the company's high performance recording/reproducing technology, developed primarily for the extreme environments of space and military defense, into the commercial arena as well as a lower-cost alternative for government and military applications previously requiring custom built instruments.

"DATATAPE has been committed to the nation's space program for several decades and we intend to keep this commitment alive by continuing the company's trend of providing the highest quality, technologically advanced products," says Saccacio. "DATATAPE is investing in technology by investing in people who will incite the next 'quantum leaps' in data recording/reproducing methodology. This has been the core competency of the company since its very inception in 1937 when Herbert Hoover, Jr., son of the 31st president of the United States, founded Consolidated Engineering Corporation — the forerunner of DATATAPE.

"DATATAPE is proud to have contributed to NASA's 35 years of space achievements and looks forward to continuing this partnership into the next generation of space exploration. We will be putting the re-engineered DATATAPE to work to meet the evolving and ever-more complex requirements of NASA."

A New Generation

DATATAPE's new generation of high-performance data acquisition and tape record/reproduce systems will continue to provide NASA, the related aerospace companies, the Department of Defense, laboratories, aircraft manufacturers, and commercial enterprises with state-of-the-art recording devices for a wide range of data-collection applications. Products introduced over the past few years include the MARS-II Multi-Application Recording System, the DTR-32 Digital Recorder/Reproducer System, and the DCTR-LP Series. Each satisfies a wide range of applications meeting the company's design philosophies of "dual-use" storage systems.

The MARS-II combines small size with multiple data formats. The system flexibility of MARS-II allows for user configuration in scaling the size of the system connecting one electronic module with up to seven storage modules.

The DTR-32 is a digital recorder with a dual-purpose capability to record instrumentation data and transfer it directly onto a computer for compilation and analysis. Thus, the DTR-32 can be used both as an instrumentation device and as a computer peripheral. This capability will be an inherent feature in future recorders as customers re-engineer the data acquisition, analysis and dissemination processes.

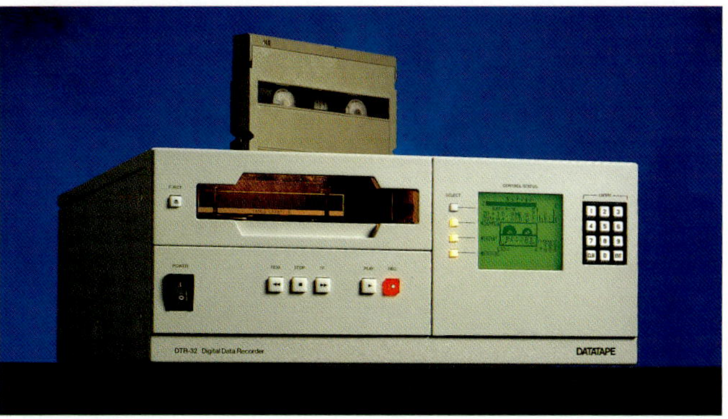

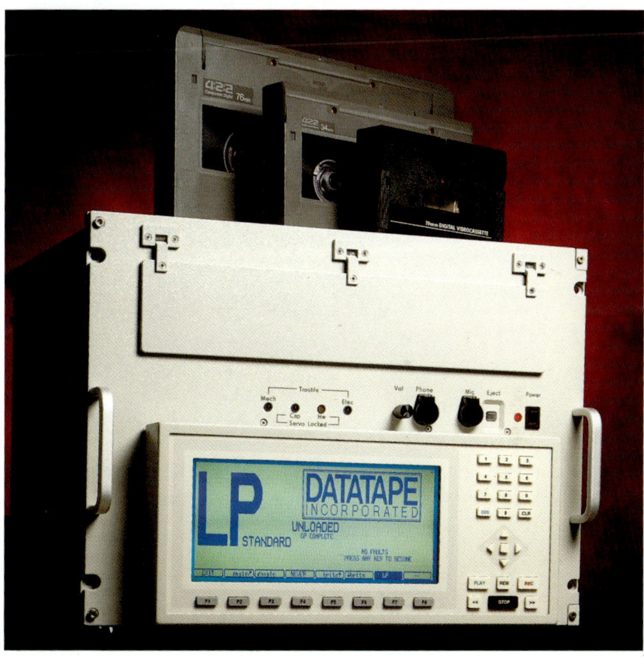

DATATAPE continues in its quest for delivering advanced technology recorder/reproducers with the recent introduction of the (top to bottom) MARS-II, DTR-32 and DCTR-LP Series.

The DCTR-LP Series offers the fastest digital recording capabilities in the world. The recorders are suitable for both data acquisition and mass-storage applications operating at transfer rates up to 400 Mbps.

DATATAPE employs more than 400 people and is headquartered in Pasadena, California, with additional operations in Santa Clara and San Diego. The company has 13 field sales, service and support offices worldwide. DATATAPE is proud to have played a vital role in America's exploration of new technological frontiers and looks forward to supporting NASA in its next generation of space exploration.

Experienced in Space Habitats

JE Supports NASA's Commitment to Space...

Johnson Engineering Corporation operates and maintains full size mockups and trainers for the Space Station and the Space Shuttle missions in Building 9N at Johnson Space Center in Houston.

Twenty-one years ago, Johnson Engineering Corporation (JE) was founded by aerospace engineer Dale R. Johnson with a commitment to supporting human ability to live safely and work productively in the harsh environment of space - to literally become "At Home In Space". JE has supported all of NASA's human spaceflight programs since Skylab in 1973 and has provided advanced concepts for future programs. An early project involved the analysis of all of the film, video and voice transcripts of the Skylab missions and the creation of a data base capturing lessons learned for use by future spacecraft designers. Today, JE continues its mission as a prime support contractor to the Flight Crew Support Division (FCSD) of NASA's Johnson Space Center. JE's responsibilities under the Flight Crew Systems Development Contract fall into three major categories:

- Crew station integration and human factors support;
- Design, development, fabrication, operations and maintenance of mockups and trainers;
- The design and development of flight hardware for use by astronauts.

CREW STATION INTEGRATION AND HUMAN FACTORS SUPPORT

Johnson Engineering is a recognized leader in the field of habitability - the process of designing spacecraft interiors which maximize the productive use of minimum volumes in carrying out living and working tasks. JE habitability engineers evolve crew accommodations which maximize crew safety, productivity, and comfort under all spaceflight conditions. JE has developed extensive expertise in the design of stowage arrangements which efficiently utilize available onboard space. JE is also responsible for producing the Crew Compartment Configuration Drawing (CCCD) which controls the stowage configuration of the crew compartment for each human flight. And, with each Space Shuttle and Space Station mission, JE provides all decals and identification placards through the decal facility located in Houston at JSC.

The Human Factors support activities ensure a safe and productive crew environment. JE specialists conduct research into crew interfaces, human performance and human capabilities and limitations in space. These studies result in guidelines, standards and recommended design solutions to deal with complex human-machine interface issues. Specialized laboratories support applied research into critical areas such as Human-Computer Interaction, Anthropometry and Biomechanics, Remote Operator Interaction, Graphics Analysis and Lighting.

MOCKUP AND TRAINER SUPPORT

Johnson Engineering supports the training of astronaut flight crews by designing, operating and maintaining the high fidelity mockups and trainers located in Building 9N at JSC.

Mockups are configured to represent the arrangement of the crew compartment for each mission. JE operates on-site fabrication facilities to craft mockups and trainers out of metal, plastic, wood, fabric, sheet metal, fiberglass and various other materials. In addition, JE modifies and refurbishes existing articles for the special requirements of each mission.

JE's responsibilities include operation of the Weightless Environment Training Facility (WETF) in Building 29 at JSC. In the WETF, astronauts train in the neutral buoyancy pools for complex Extra Vehicular Activity (EVA) tasks in space. The effectiveness of this training environment has been demonstrated by the success of such missions as the Intelsat capture and the Hubble Telescope repair. With each mission, JE is responsible for the design and fabrication of all the mockups and trainers used underwater for training astronauts.

In addition to the underwater training facilities, the Precision Air Bearing Facility (PABF) and the Partial Gravity Simulator (PGS) are available to simulate specific aspects of the reduced gravity environment for astronauts.

FLIGHT HARDWARE DEVELOPMENT

Being at home in space depends on having the right equipment and provisions on hand for every need. JE is responsible for providing these necessities including flight crew equipment, tools, food, clothing, personal hygiene provisions, cameras, experiments and payload related systems. This function encompasses the tasks of creating concepts, performing engineering studies and technical analyses, designing components and subsystems, production of prototypes and test articles, manufacture of flight hardware, testing and certification of components and subsystems, interface definition and implementation, production and maintenance of documentation, configuration management, mission logistics support, sustaining engineering, subsystem management support, realtime mission support and post-mission analysis of inflight performance.

While in space, the comfort and safety of the flight crew remains a primary focus and responsibility of JE. As the space program evolves into the 21st century and NASA looks for ways to be "at home in space", JE will be there supporting the next human adventure.

The underwater trainers for Space Shuttle and Space Station are in the Weightless Environment Training Facility (WETF) in Building 29.

A rendering of the future Neutral Buoyancy Laboratory (NBL) being built for the underwater training of astronauts.

JE designed and built a hand-operated trash compactor that reduces waste volume 8 to 1 for use on long duration Space Shuttle missions.

AlliedSignal Aerospace

Future missions of the NASA Space Station may use solar generators as part of a Solar Dynamic Power System made by AlliedSignal Aerospace that will generate electric power by a closed Brayton Cycle. Designed to operate 24 hours a day for up to 30 years, the system's modular design will permit increases in electrical power to meet future requirements.

AlliedSignal Aerospace's experience in environmental control/life support and thermal control systems, gained through NASA's Mercury, Gemini and Apollo programs, will ensure that the Space Station maintains a safe and comfortable working and living environment for our astronauts in the vacuum of space. AlliedSignal has been selected to provide subsystems and assemblies for carbon dioxide removal, internal thermal control, fire detection and suppression, valves and the berthing mechanism to help assemble the Space Station in Earth orbit. NASA's European and Japanese partners also have selected AlliedSignal Aerospace to help make the Space Station an international program.

An AlliedSignal Aerospace Control Movement Gyro System will supply the attitude control actuation for the Space Station. AlliedSignal Aerospace provided a similar control concept for NASA's Skylab spacecraft. AlliedSignal Aerospace's control and navigation systems were also used by the Saturn rocket, which launched early U.S. spacecraft.

With these projects, AlliedSignal Aerospace, one of the world's largest manufacturers and suppliers of advanced technologies and technical services for space, military, commercial and general aviation, is continuing 65 years of technical excellence that enables man to live and work in space. Among its businesses are trade names from the pioneering days of powered flight — Bendix, Garrett and AiResearch. These businesses have supplied essential components for nearly every type of civil and military aircraft and spacecraft in the West.

AlliedSignal Technical Services Corporation's employees assist in the tracking of the space shuttle in the Mission Control Room at Johnson Space Center, Houston, Texas.

The Goldstone 26-meter tracking antenna is used for backup support for all STS missions, manned flight support, and near-orbit spacecraft. AlliedSignal Technical Services Corporation's engineers and technicians operate the antenna.

AlliedSignal Aerospace's contribution to the space program consists of more than flight hardware and systems. The company's Technical Services Corporation in Columbia, Md., has been a world leader in spacecraft mission operations for nearly 40 years. These operations include tracking, communications, data acquisition, information processing and command and control. AlliedSignal has performed these services for NASA and other government agencies continuously since 1958, supporting every NASA scientific and manned spacecraft mission from Project Vanguard to the present day Space Shuttle. The company supports NASA's Goddard Space Flight Center, where one of its functions is to manage, operate and maintain the second Tracking and Data Relay Satellites' (TDRSS) Ground Terminal, which links and controls the geosynchronous TDRSS with Earth resources and users; the Johnson Space Center, where it is responsible for the maintenance and operations, sustaining engineering, and ground support segment of direct mission operations functions for the Shuttle Operations Contract; the Ames Research Center, where it provides data processing and flight operations support for the Pioneer program; the Jet Propulsion Laboratory, where it operates and maintains NASA's Deep Space Network and provides the link with deep space probes on such missions as Voyager, Magellan and Galileo; and the Lewis Research Center, where the Brayton Rotating Unit of the closed Brayton cycle system has been successfully tested.

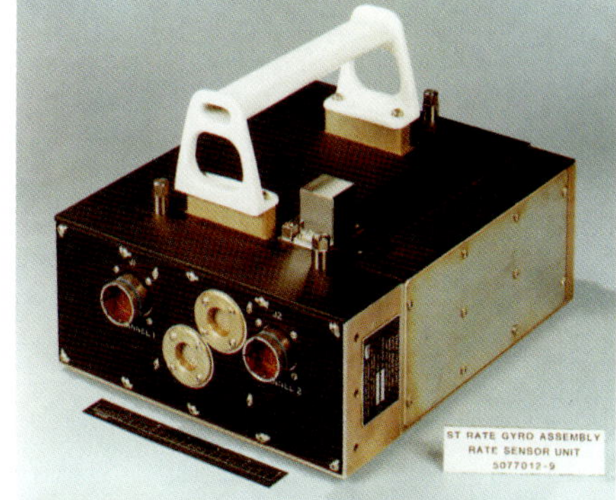

The Control Movement Gyro System for the Space Station (left) is similar to a control concept that AlliedSignal supplied for NASA's Skylab spacecraft. For the Hubble Space Telescope, AlliedSignal Aerospace provides a pointing control system that includes the Gyro Electronics Control Unit (center) and the Rate Gyro Assembly (right).

In the 1940s, AlliedSignal's Aerospace Equipment Systems developed the first pressurization and environmental control systems for military, then commercial aircraft. The company designed and built pressurization and cooling equipment for the X-15, which penetrated outer space, enabling pilots to leave the Earth's atmosphere for the first time. AlliedSignal's system for the X-15 pressurized and ventilated the pilot's flight suit, cooled electronic equipment and the plane's nose cone and ram inlet, and operated pneumatic equipment. Since then, AlliedSignal's environmental control systems have maintained astronauts throughout their epic missions, both inside and outside the spacecraft. The company developed Extravehicular Activity (EVA) life support systems used during spacewalks. From Project Mercury through Gemini, Apollo, Skylab, and Apollo-Soyuz, AlliedSignal has provided the vital life support systems to keep men alive and comfortable in space.

In addition to environmental control systems for the Apollo command module, AlliedSignal Aerospace provided the Apollo Lunar Surface Experiments Package, six scientific stations left on the moon to form an automated network of lunar science laboratories. Designed to work only one year, the Apollo 11 ALSEP system transmitted an uninterrupted stream of scientific and engineering data from the moon for more than five years. The lunar roving vehicle utilized AlliedSignal electric propulsion motors for its four wheels and for power steering during moon exploration and experiments.

Skylab utilized AlliedSignal-manufactured astronaut life-support assembly, regenerative carbon dioxide removal system, heat exchangers, fans, control valves, and chiller system.

For the space shuttle, AlliedSignal Aerospace makes a digital transducer assembly which senses basic air data parameters from the shuttle's pitot-static system and converts them into highly accurate digital data for use by the orbiter's general purpose computer to calculate vital air data, such as altitude, angle of attack and air speed during reentry. The company supplies the Solid Rocket Booster (SRB) integrated electronics assembly (IEA), which serves as the only electrical interface between the orbiter and the SRB systems. It supplies the signals to ignite the two booster rockets and is instrumental in controlling the flight path of the space shuttle during SRB burn. The assembly also commands explosive charges which separate the SRBs from the space shuttle. Once separation is complete, the IEA relies on its own electronic logic to control all phases of the SRB recovery.

AlliedSignal Aerospace produced three instruments for the shuttle orbiter: a surface position indicator, which displays essential flight parameters to the crew, and two additional vertical scale instruments, which provide essential flight information.

For Hubble Space Telescope, AlliedSignal Aerospace provided a pointing control system consisting of rate gyros, star trackers, reaction wheels, a magnetic torquing system and safe mode control electronics, as well as the control laws and algorithms for pointing, star scanning, planet tracking and spacecraft maneuvering. An AlliedSignal Aerospace star scanner is also aboard Galileo.

AlliedSignal Aerospace continues research and development on numerous projects that will have a bearing on future space activities. Among them: a laser gyro-based navigation and control system design for spacecraft and launch vehicles; a Sabatier reactor life-support system for long-duration space activities; environmental control and life-support systems for a space station operating continuously for 10 years; advanced delivery concept for orbital resupply of space vehicles; tests of astronaut capabilities to perform work at zero gravity; production of a hypersonic ramjet engine using liquid hydrogen fuels and designed for flight speeds between Mach 3 and Mach 8; and rotary actuators for the high-lift actuation system in a super-critical wing foil project.

AlliedSignal Aerospace is organized into four major businesses — Aerospace Equipment Systems, Commercial Avionics Systems, Government Electronic Systems and Engines — comprising one of the largest aerospace equipment manufacturers in the world. AlliedSignal Aerospace, along with AlliedSignal Automotive and AlliedSignal Engineered Materials, make up AlliedSignal Inc., an advanced technology and manufacturing company serving customers worldwide with aerospace and automotive products, chemicals, fibers, plastics and advanced materials.

Space Vector Corporation

Since its founding in 1969, Space Vector Corporation (SVC) of Chatsworth, California, has contributed to expanding our access to space by providing launch vehicles and launch services for scientific and military missions.

Today, SVC continues its important role of providing launch services to the international scientific community for suborbital missions, launch and payload integration services to the Department of Defense (DoD) and the development of low cost satellite launch vehicles for Low Earth Orbit (LEO) operations. SVC's spacecraft, subsystems and components are the standard for sub-orbital missions.

In-House Capability

SVC maintains substantial on-site engineering, manufacturing, test and integration capability at its facility in Chatsworth, CA. These capabilities provide a "total system approach" from the conceptual design phase through to final test and evaluation.

SVC's computerized design, analysis and simulation capabilities support a variety of space science applications, including launch vehicle and spacecraft trajectory simulations; F.E.A. tools for structures and thermodynamics; and specialized systems for analog/digital micro-controlled circuit design and test.

The manufacturing facility houses a complete production line of aerospace machine tools and maintains a prototype line equipped for "short run," unique or specialty items. Spacecraft electronic components are assembled on a dedicated production line at the SVC product development facility. SVC's test and integration capabilities include: a "free-flight" hardware-in-the-loop test facility, automated multi-purpose electronic test stations, vehicle system integration and test facility, environmental test stations and guidance and control (GNC) development laboratories, and clean room(s).

Launch Vehicle Systems

Among the systems and components SVC has developed for space applications are exoatmospheric control, specialized inertial and electronic devices, thruster systems and parachute recovery systems. Each vehicle subsystem is available as an individual component or as an element of an integrated flight module.

Exoatmospheric control systems developed by SVC for sounding rocket applications provide the science customer with mission performance capability. Users include: U.S. government agencies including NASA, Army, Air Force and Navy; foreign government agencies including Germany, Canada, and Brazil; colleges and universities; national research laboratories, and private corporations.

SVC inertial sensor devices include the Midas Platform, a roll stabilized platform; Rate Control System, a zero-gee control package; Magnetic Control System, a "gyro-less" system sensing the Earth's magnetic fields; and the Advanced Guidance System (AGS), an integrated "strap-down" guidance system.

The AGS is specifically designed for missions requiring attitude and positional reference. It is ideal for applications requiring boost control and high precision positioning. Range safety destruct criterion can be provided from real-time AGS reference data for launches from remote sites without extensive FTS resources.

SVC produces cold-gas thruster devices with features unique in the industry. The Reaction Control System (RCS) is adjustable, capable of supplying thruster forces from 0.1 to 25 lbf. This system is suited to applications including sounding rocket attitude control systems, control of space platforms, sensor modules and satellites. The company also provides a variety of mechanical and pneumatic separation and release mechanisms.

Mission Integration and Launch

As mission integrator for suborbital science experiments, SVC provides the full scope of services required to support space-based science programs, from mission concept development through hardware fabrication and test to post flight analysis. Specialized services include payload recovery, zero gravity tests, video links, solar/star alignment and man-in-the-loop control.

The Aries sounding rocket series, developed by SVC, is based on Minuteman ICBM upper stage rocket motors. The Aries provides suborbital mission trajectories for experiments with large diameter and/or heavy payloads. The Aries can be launched either from a government launch range or from launch control ground support equipment erected in a remote area. The first commercial space launch vehicle (the Aries-Conestoga) was designed, fabricated, deployed and launched by SVC under contract with SSI of Texas, from an unimproved site on the island Matagorda in the Gulf of Mexico.

Using the innovative SVC approach to space flight systems, small satellite launches are achievable at a fraction of the cost of "traditional" approaches. A launch vehicle utilizing a combination of commercially available and surplus strategic assets can perform satellite injection into LEO with significant cost savings. For example, programs can take advantage of the significant cost savings available through the utilization of government surplus ICBM rocket motors. Minuteman motor configurations are capable of placing payloads of approximately 500 pounds into LEO.

DoD Space Technology Programs

As a major Department of Defense subcontractor for space technology programs, SVC provides military space flight test support. SVC DoD vehicle configurations include the target vehicle for U.S. Army Theater Missile Defense system, the flight test vehicle for interceptor sensor demonstration efforts and the launch vehicle for ground based interceptor demonstrations. Specialized launch vehicle features developed to support DoD mission requirements include: clam-shell nose fairings, multi-staged target vehicles, real time video down-links, "man-in-the-loop" attitude control systems, independent LV and spacecraft guidance, portable launch control ground support equipment and mobile vehicle transporter/erector/launchers.

SVC is the recipient of several U.S. Army Strategic Defense Command Technical Achievement Awards, including those in 1983 for the Homing Overlay Experiment; 1986 for SDIO – Delta 180 Target Vehicle, and in 1992 for the Exoatmospheric RV Interceptor System (ERIS).

Capabilities from the Past—Applications for the Future

A remarkable 25-year history of developing reliable launch vehicle products, with strong emphasis on quality and responsiveness, has earned SVC a place as an industry leader in the sub-orbital aerospace segment. Recent advances in technology, including the AGS and the SVC Universal Upper Stage (USS) are critical in SVC's drive toward the deployment of new small satellite launcher systems. Upon this base, SVC works toward its goal of an ever expanding access to space.

Aerotherm Corporation

As an early partner in the Apollo, Gemini, and deep-space programs, Aerotherm supported NASA with hyperthermal test capabilities and material response expertise. For more than 30 years, Aerotherm has been testing and analyzing materials to determine their response to the extreme environments encountered during reentry into the Earth's atmosphere. Space shuttle tiles were fired to reentry temperatures in Aerotherm's arc plasma laboratory, which is located adjacent to NASA Ames. Aerotherm also developed a thermal protection material repair kit for damaged insulation on the space shuttle's external fuel tank. For NASA's Solid Propulsion Integrity Program, Aerotherm helped generate a computer code for analyzing the performance of solid-rocket motor nozzles. Aerotherm has also developed hybrid- and liquid-rocket engine concepts supporting NASA's pursuit of safer, lower cost, higher performance rocket engines.

Building on this early aerospace work, Aerotherm has become a leading supplier of products and services in the following areas: computational fluid dynamics, structural and material analysis, laser effects, solid-rocket propulsion systems, systems engineering of signature-controlled (RF and IR) space and missile payloads, and test and evaluation operations support.

Aerospace Systems Engineering

Since its founding in 1965, Aerotherm has supported NASA with systems engineering and aerospace materials technology studies. Materials performance models developed at Aerotherm are in use at all major aerospace systems companies, NASA centers, and DOD laboratories.

Aerotherm has designed, fabricated, and installed arc plasma test facilities at NASA Langley Research Center, NASA Johnson Space Center, and NASA Ames Research Center, as well as at several DOD and contractor research centers.

Recently, Aerotherm was chosen as a team member on the NASA-sponsored COMET program, which is designed to provide low-cost recoverable access to space for microgravity experiments. Aerotherm is providing the thermal protection system for the COMET capsule.

In addition, Aerotherm designs and flight tests high-temperature, signature-controlled payloads. Aerotherm personnel are experts in ICBM strategic and theater missile defense (TMD) countermeasures, active and passive RF/IR payloads, signature analysis, discrimination, and system effectiveness. A variety of target payloads for the Ballistic Missile Defense Organization's TMD Programs are manufactured by Aerotherm.

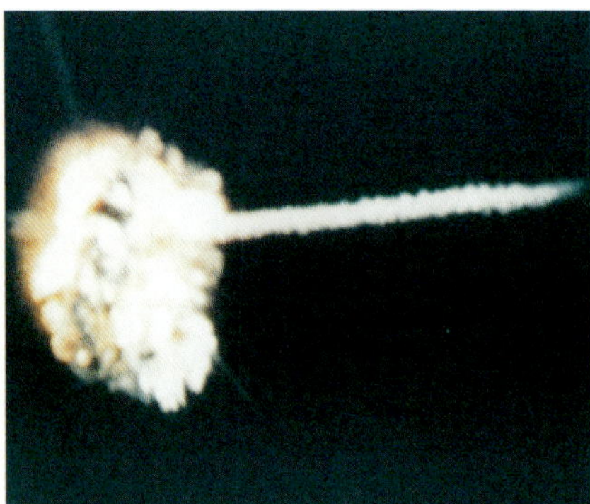

An ERINT fired from a Patriot launcher tracks and destroys a Storm target reentry vehicle made by Aerotherm.

International Programs

Aerotherm expertise is also supporting international aerospace programs. For example, Aerotherm is designing the thermal protection system (TPS) for Italy's CARINA space capsule, which will provide recoverable access to space for microgravity experiments. Aerotherm, under license, will transfer the manufacturing technology to Alenia Spazio to allow fabrication of the CARINA TPS in Italy.

The COMET capsule, part of a NASA-sponsored program to provide access to space for commercial experiments, will be protected during reentry by a thermal protection system designed and fabricated by Aerotherm Corporation.

Engineers at Aerotherm recently completed the design of the plasma wind tunnel (PWT) test facility to be installed at the Centro Italiano Ricerche Aerospaziali. Jointly sponsored by Italy, the European Space Agency, and the Centre National d'Etudes Spatiales, fabrication of the PWT is scheduled to begin in 1995. When completed, this 70-MW PWT will be the largest such facility in the world. In addition, Aerotherm's inhouse arc plasma facility was used to test candidate thermal protection materials for the European Space Agency's Hermes and Japanese aerospace programs.

Aerotherm arc plasma heaters are used to test components for space transportation systems, such as the U.S. Shuttle, ESA's Hermes, and Japan's HOPE.

These examples illustrate Aerotherm's support of NASA's commercialization and technology transfer programs to benefit the United States and other nations.

Technical and Engineering Services

In response to our nation's need for increased productivity in our NASA and DOD operations, Aerotherm's Technical and Engineering Services organization provides a complete spectrum of capabilities for operation and maintenance of NASA, DOD, and DOE laboratories and test facilities. For example, in New Mexico, Aerotherm operates the High Energy Laser Systems Test Facility (HELSTF) for the U.S. Army Space and Strategic Defense Command. This facility includes one of the most powerful chemical lasers (MIRACL) in the world, as well as a beam transfer center, which directs the MIRACL beam to hazardous testing, vacuum testing, or effects testing areas. In addition, the Navy's SEALITE beam director can focus the MIRACL power on dynamic airborne or ground-based targets. HELSTF's pulsed laser vulnerability system can assess the susceptibility and vulnerability of U.S. tactical weapons against postulated directed energy threats.

Under Aerotherm's leadership, the HELSTF testing schedule has been maintained while using 35 percent fewer people. The ability to manage complex test facilities, increase productivity, and meet tight budget constraints is a hallmark of Aerotherm's relationship with the government sector. At HELSTF and other advanced technology facilities, Aerotherm is providing the individualized, quality services necessary for highly successful and cost-effective operation.

As the next century begins, Aerotherm Corporation looks forward to supporting NASA in upcoming space missions as well as down-to-earth facility operations.

The High Energy Laser System Test facility, managed by Aerotherm, demonstrates how increases in productivity and reductions in costs can be accomplished at government installations. This photo shows the high-speed shutters and turning mirrors that direct MIRACL, one of the world's largest chemical lasers, to various test areas.

MANAGEMENT TECHNOLOGY NSI

Managing Space Technology

ManTech International Corporation is proud to partner with NASA in making it possible to be "at home in space." Technical know-how, preparedness, and problem solving are key to NASA's successful scientific exploration. ManTech supports NASA in this endeavor by delivering high technology systems which are innovative and effective.

This is ManTech ...

ManTech, headquartered in Fairfax, Virginia, is a major professional and technical services company. We number over 3,500 professionals in over 100 worldwide locations. Our subsidiary, NSI Technology Services Corporation, has two operation centers: Applied Technology Center (ATC) and Aerospace Technology Applications Center (ATAC). These centers perform multi-disciplined engineering, design, fabrication, assembly, integration, operation, maintenance, and test and evaluation functions in support of national priorities within NASA.

For over 24 years, NSI has provided support services on many scientific NASA projects such as the Kuiper Airborne Observatory (KAO), flight simulators and the Hubble Space Telescope (HST) First Servicing Mission. At Goddard Space Flight Center, NSI has provided test engineering and operations support for every major spacecraft program since 1971. Recent ongoing work includes prime responsibility for developing thermal experiment payloads for Space Station and Earth Observing System (EOS) applications.

In flight, NSI operates the KAO telescope and data systems.

NSI Support of the Kuiper Airborne Observatory

NSI has supported the operation of the NASA-Ames Kuiper Airborne Observatory since 1978. We provide engineering, technical, expedition, and logistics support for approximately 80 research missions per year. In flight, NSI personnel operate the telescope, star tracker, and data systems. With NSI's support, astronomical research scientists have explored supernovas and the rings around Uranus, confirmed the atmosphere on Pluto, and observed the Shoemaker-Levy 9 Comet.

NSI Aerospace and Human Factors Research Simulation Support

Also at Ames, NSI maintained, operated, and provided logistics support for two sophisticated aerospace flight simulators. One, the Vertical Motion Simulator, was configured with the Shuttle Cab so that NASA astronauts could practice the flight dynamics of the Shuttle reentering the Earth's atmosphere. A second, the Crew Vehicle Systems Research Facility, provided a broad-based, flexible testbed for human factors research. A fully functional Boeing 747 simulator and an Advanced Concepts Cockpit provided out-of-window imagery during flight simulation operations.

NSI Support of the Hubble Space Telescope First Servicing Mission

Launched in 1990, the HST, a national resource, is a powerful astronomical observatory designed to be serviced on-orbit. NSI at Goddard supported NASA in its three mission objectives: correct the imperfect optics, restore reliability of the spacecraft, and prove the viability of on-orbit servicing. This task followed NSI's earlier environmental test support for NASA on the Solar Maximum Repair Mission (STS-41C in 1984).

For the HST mission (STS-61 in December 1993), NSI supported the preparation of systems needed to replace parts on the telescope. The NSI team provided mechanical integration of the Instrument and Solar

NSI crews optically measured the Wide Field Planetary Camera-II location in the HFMS.

Array Carriers, and supported in-house testing of the instruments and flight avionics. Other activities included thermal blanket preparation and electrical harness fabrication. Our personnel provided optical alignment measurements, contamination control, and safety engineering at Goddard and the Kennedy launch site.

The NSI team developed, assembled, and optically aligned a full scale replica of the flight telescope optical bench. We designed and built the Vehicle Electrical Systems Test (VEST), an exact mockup of the HST Equipment Section and Optical Test Assembly electrical equipment bays. Also, NSI fabricated electrical harnesses for and assembled the High Fidelity Mechanical Simulator (HFMS).

The VEST and HFMS were designed and built to perform functional and fit checks of HST flight instruments and avionics. These simulators provided the flight crew with a highly accurate means to develop procedures, gain handling experience, and rehearse techniques which would be needed for on-orbit servicing. During the mission, ground crews, supported by the NSI integration team, manned these simulators to resolve any unexpected difficulties encountered by the astronauts in space.

Special equipment was needed to handle and integrate flight instruments such as the Wide Field Planetary Camera-II and Corrective Optics Space Telescope Axial Replacement. NSI structural engineers designed and built the innovative Vertical Integration Facility. This sophisticated payload handling facility was designed to prevent potential damage to critical interfaces during latching maneuvers.

Environmental Testing

NSI personnel maintained and operated all of the environmental test facilities needed to qualify the HST flight hardware. Acoustic, EMI, magnetic, high capacity centrifuge, static load, vibration, shock, and thermal vacuum testing were conducted. Our structural engineers devised a first ever "twang" test in the Modal Facility to excite high-deflection launch and landing loads on the Orbital Replaceable Unit (ORU) and Solar Array Carrier.

Weightlessness Simulation/Handling

NSI formed a special team of SCUBA divers to support NASA in underwater, neutral buoyancy testing facilities. Our divers assisted the project team assemble flight mockups underwater where they could simulate weightless effects of EVA operations in space.

Handling crews of NSI and other personnel supported all logistics in transporting flight hardware between NASA centers and the Kennedy launch site. The main flight instrument carriers were shipped via truck and ocean-going barge from Goddard to the launch site. At Kennedy, NSI integration crews readied the flight hardware for installation into the Shuttle and removed equipment returned from space after the mission.

NSI performed mechanical integration on the ORU Carrier in the Vertical Integration Facility.

HST Servicing Mission Success

By February 1994, enthusiastic scientists around the world were praising the telescope's substantially improved images. This was demonstrated visually by the images of the Shoemaker-Levy 9 Comet's collision with Jupiter in July 1994. Meanwhile, NSI is already supporting NASA's preparations for the HST Second Servicing Mission scheduled for 1997.

BOOZ·ALLEN & HAMILTON

A Rich History of Contributions to the US Space Program

Booz·Allen & Hamilton has served NASA since 1963 on programs ranging from Space Station engineering and integration, to manned and unmanned spaceflight operations, to astronaut training, to communications network support. The firm's services span the life cycle of a space system — from helping define requirements and identify the most effective initial concepts through development, production, operation, support, and exploitation of the information gathered. At Booz·Allen, engineers, scientists, and computer systems specialists work side-by-side with senior space experts who have hands-on experience in manned space flight, shuttle operations and support, satellite systems and information networks.

This computer-aided design model was created by Booz·Allen staff supporting NASA in Space Station assembly planning.

Technology and Management Support for the Space Program

Astronauts assembling a space station in orbit above Earth...remote sensors exploring distant planets...and launch vehicles lifting precious communications satellites into position all rely on the same lifeline — the critical Spaceflight Tracking and Data Network (STDN). Because STDN's reliability must be flawless, Booz·Allen & Hamilton has developed software and simulator systems which ensure the utmost reliability and efficiency in spacecraft tracking, communications, and data links.

This is just one important element of Booz·Allen's network and mission operations support for the U.S. space program. The international management and technology consulting firm is providing systems engineering, program management, independent verification and validation, and software engineering support to NASA's Goddard Space Flight Center, Johnson Space Center and to the international partners of the European Space Agency and the Canadian Space Agency.

Booz·Allen has provided extensive systems engineering and integration support to the Space Station, including development of program technical requirements, development and analysis of top-level schedule, independent technical assessments and engineering oversight, and monitoring and assessment of engineering activities. The firm provides operations cost assessment and strategic level operations and utilization planning for the Space Station at NASA Headquarters and the Johnson Space Center. Together with CAE-Link, Booz·Allen supports the development of crew training simulators for the Space Shuttle, Space Station, and Spacelab. Separately, the firm has developed a PC-based system simulator for satellite operator training.

Booz·Allen has supported NASA's Space Station international cooperative activities with negotiation and maintenance of joint documentation, executive support for joint management mechanisms, monitoring of international partner program activities, evaluation of international agreements, analysis of shared operations costs, and strategic planning and technology transfer policy support.

As NASA looks to the future, Booz·Allen is playing a key role for the new Mission Control Center at NASA Johnson Space Center in Houston. The firm provides systems engineering support for the hardware, software, and network architecture for this new center which will provide flight planning and ground command and control of the Space Shuttle, the future International Space Station Alpha, and other space vehicles, including Russian vehicles. Among Booz·Allen's responsibilities are the design of the data architecture; the definition of engineering strategies, processes, and standards; the design and integration of the Vehicle Master Data Base; and chairing the Mission Systems Contract process improvement team.

International, Defense and Commercial Space

Government agencies and aerospace companies worldwide turn to Booz·Allen for expert, unbiased assistance in civilian, commercial, and defense space programs. Booz·Allen has served the European Space Agency and French Space Agency, CNES, in key program areas such as Columbus, remote sensing, and microgravity research, and supports the Canadian Space Agency in the development of its Mobile Servicing System for use on the International Space Station.

In the defense arena, the firm has lead roles in major defense space efforts such as the Strategic

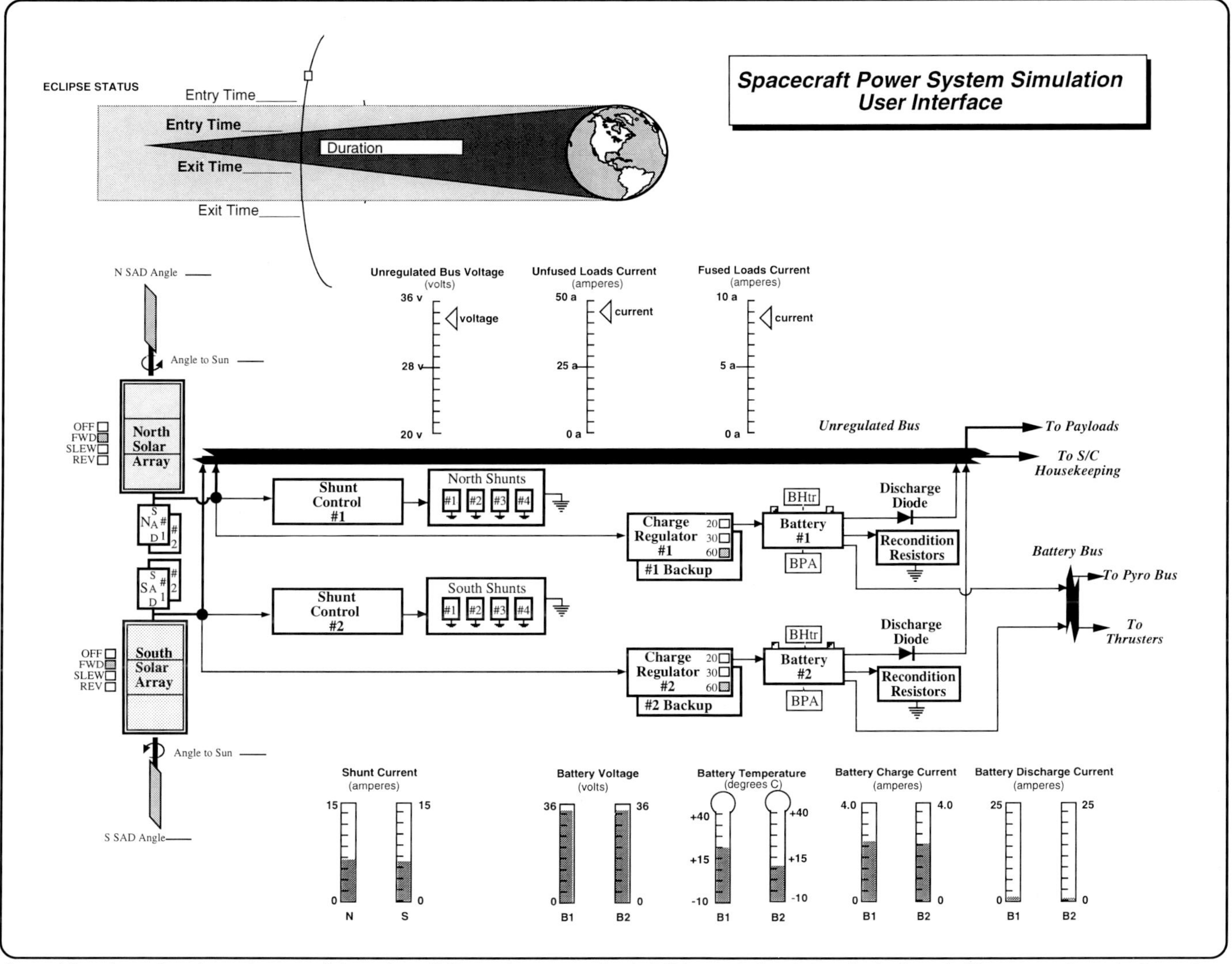

Booz·Allen has developed powerful and realistic, yet economical, PC-based simulators to train satellite ground controllers. For example, the interactive display screen enables student controllers to learn how to monitor and control critical power systems.

Defense Initiative, Milstar, and US military satellite programs, providing expert services in technology development, simulation, and communications and ground systems support. Booz·Allen also performs space systems analysis and provides engineering and technical support for the US Navy, Air Force Space Command, and Air Force Space Systems Division.

By applying a proven record of successful business management, an international business presence, an array of space technology skills, and a commitment to the entrepreneurial spirit, Booz·Allen has built a strong business base in the commercial space development arena. The firm provides commercial market research and evaluation, strategic planning for space venture financing and program development, and market-oriented systems engineering for an international roster of space business clients. For example, the firm has performed an assignment for a large European aerospace firm to study the market potential for Light Sats in defense, commercial, and academic applications over a projected ten-year horizon. The study scope encompassed European, U.S., and other developing world markets.

About the Firm

Founded in 1914, Booz·Allen & Hamilton is one of the world's largest management and technology consulting firms. Today, Booz·Allen provides services in strategy, operations, systems, and technology to clients on a global scale. Clients include most of the world's largest industrial companies and financial services firms, virtually all of the departments and agencies of the U.S. federal government, and major institutions and government bodies across the globe. Booz·Allen's Worldwide Technology Business provides technology, engineering, and management services as well as systems development and systems integration services in the areas of defense and intelligence, transportation and space, information systems, telecommunications, energy and environment.

In addition to its three-decades of service to NASA and international space programs, Booz·Allen is actively involved in the transfer of space technologies to solve earth-based challenges in areas such as transportation through development of vessel tracking, air traffic control, and intelligent vehicle systems.

SPACEHAB®

The practical benefits of the space program are being brought down to Earth by SPACEHAB. Astronauts working in SPACEHAB Space Research Laboratories aboard orbiting space shuttles conduct scientific experiments that could result in new technologies and products for the marketplace. Commercial research performed in the laboratories is preparing U.S. industry for future use of International Space Station *Alpha*.

Pioneered by SPACEHAB, Inc., the Space Research Laboratory is the world's first privately developed research and development facility that can support humans in space. Research conducted in SPACEHAB laboratories is generating revolutionary advances in biotechnology, advanced materials and manufacturing technologies that promise marketability on Earth and provide participating companies with a global competitive advantage. More than 60 U.S. companies have conducted product research on SPACEHAB flights since 1993. A sampling of possible products and technologies resulting from experiments conducted in SPACEHAB laboratories include: a new time-release medicine for diabetes; treatments for debilitating diseases such as cancer, osteoporosis, AIDS and other immune diseases; materials for longer-duration, extended-wear contact lenses; higher-yielding, disease-resistant crops; innovative robot services; improved computer chips; and harder, stronger industrial tools.

SPACEHAB, shown here in *Discovery's* open payload bay, doubles the available living and working space on the shuttle for astronauts and quadruples the available experimentation space. (Photo: NASA.)

SPACEHAB, Inc. is a privately held corporation established in 1983 to design, build and lease pressurized Space Research Laboratories that fly aboard NASA's space shuttles. These orbital laboratories enhance shuttle operations to encourage the commercial use of space by U.S. industry. SPACEHAB is the first company engaged solely in the business of providing frequent, affordable access to an astronaut-tended environment in space.

The patented SPACEHAB laboratories are pressurized, cylindrical modules that measure 10 feet in length by 13.5 feet in diameter, with a truncated top and flat "end-caps." During a typical flight in the space shuttle, the laboratory is located in the forward quarter of the payload bay and connects to the astronaut compartment through the space shuttle airlock by a short tunnel. SPACEHAB doubles the available living and working space on the shuttle for astronauts and quadruples the available experimentation space.

The SPACEHAB Space Research Laboratory made its maiden voyage aboard space shuttle *Endeavour* during June-July 1993. The 22 experiments con-

Astronauts Janice E. Voss (foreground), Brian Duffy and Nancy J. Sherlock performed experiments during the SPACEHAB Space Research Laboratory's maiden voyage, June 21-July 1, 1993. (Photo: NASA.)

Astronauts Ronald M. Sega and N. Jan Davis performed experiments in the SPACEHAB Space Research Laboratory aboard *Discovery* in February 1994. (Photo: NASA.)

ducted onboard were sponsored by more than 25 U.S. companies, NASA, a network of universities and research institutions, and the European Space Agency. Astronaut David Low, SPACEHAB payload commander, said the new lab "performed like a champ."

Among the experiments was the growth of high-quality Malic Enzyme crystals in an effort to determine the protein's molecular structure. This knowledge could help scientists develop a disease-fighting agent to stop parasite infections in humans. With such a treatment available, there could be a significant reduction in parasite infections that currently affect one-third of the population in third-world countries.

Among the 13 experiments conducted in SPACEHAB during its second mission aboard space shuttle *Discovery* in February 1994 was the space testing of a new drug that could be used to treat AIDS symptoms and other immune deficiencies.

Another experiment grew crystals of human insulin that will be studied using x-rays to see if their structure reveals clues to aid development of a new, longer-lasting treatment for diabetes. Scientists defined a previously unknown molecular site on the human insulin protein crystals grown in space which could bond with a time-release drug for a treatment that could reduce the number of painful injections required by the 14 million Americans who suffer from diabetes.

Also during SPACEHAB's second flight, the Space Experiment Facility grew a crystal of cadmium telluride inside its 1100°F furnace. The crystal will be used in the development of improved semiconductors for radiation detectors.

Polymers (thin-film filters) grown in space during SPACEHAB's first and third voyages will be used to manufacture contact lenses superior to those produced on Earth. Polymers grown in space have improved gas permeability and durability over those grown simultaneously on Earth.

As space shuttle *Discovery* rendezvoused with the Russian Space Station *Mir* in February 1995, astronauts onboard conducted 21 scientific and commercial investigations inside SPACEHAB, which was making its third flight into orbit. Included were 12 biotechnology experiments, ranging from plant growth to pharmaceutical development; three advanced materials experiments, such as contact lens material development and metal mixing to create stronger alloys; four technology demonstrations, including on-orbit robotic experiment servicing and radiation monitoring; and two pieces of supporting hardware measuring on-orbit accelerations. Helping astronauts perform experiment operation tasks was a new robot named Charlotte™. The cable-suspended robot practiced inspection and manipulation tasks that will save astronauts time in space, potentially on International Space Station *Alpha*.

An artist's rendering of a U.S. space shuttle docked with the Russian Space Station *Mir*. (Copyright SPACEHAB.)

SPACEHAB's Payload Processing Facility, where scientists prepare their experiments for flight, is equipped with a 100,000 class clean room. (Photo: SPACEHAB.)

The applications research being conducted on SPACEHAB today is building a bridge toward tomorrow's International Space Station *Alpha*. SPACEHAB is negotiating with NASA for the use of its new "double-module" laboratories in space shuttles on trips to the Russian Space Station *Mir*. These would provide twice as much volume to conduct scientific and commercial experiments and to carry supplies to and from *Mir*. SPACEHAB also is available to carry experiments and supplies on space shuttles during the assembly phases of International Space Station *Alpha* through 2002 and beyond.

Headquartered in Arlington, Virginia, SPACEHAB is managed by a small entrepreneurial staff with experience in venture financing, in space travel as astronauts, and in large space program management in the aerospace industry and NASA. McDonnell Douglas Aerospace, Huntsville, Alabama, is SPACEHAB's prime contractor, employing more than 100 people in Huntsville and at the SPACEHAB Payload Processing Facility in Port Canaveral, Florida. SPACEHAB is represented in Europe by INTOSPACE, and in Japan by Mitsubishi Corporation, for marketing experiment space aboard the laboratories.

Evans & Sutherland

When astronauts begin building the space station, and later live and work there, they will perform tasks they have practiced many times over using high-end, real-time computer graphics developed by Evans & Sutherland.

Evans & Sutherland's image generators have aided in astronaut training since 1976, when the company provided NASA a CT3 image generator. The CT6 system, an improved multi-function scene generation system installed at Johnson Space Center in 1988, aided astronauts training to perform space shuttle orbiter operations, payload handling, remote manipulator operations, space walks, space station construction and operation, and docking and berthing techniques.

Astronaut training and solutions to aerospace engineering problems are now being provided with Evans & Sutherland's ESIG-3000 visual systems installed at the Johnson Space Center Shuttle Mission Training Facility and Systems Engineering Simulator. Evans & Sutherland's high-end visual systems also are aiding researchers at NASA Ames Research Center and NASA Langley Research Center, as well as aerospace companies.

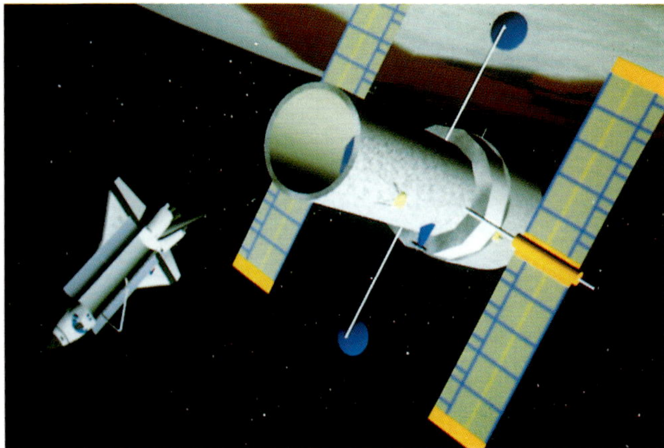

This "out-the-window" computer generated view of a satellite being deployed from space shuttle aids in astronaut training. (ESIG-3000.)

Leading Edge Computer Graphics

Evans & Sutherland was founded in 1968 by two pioneers in the field of computer science, Dr. David C. Evans and Dr. Ivan E. Sutherland, who joined forces to develop the technology that would allow computers to be used as powerful tools. Since that time, the company has been at the leading edge of computer graphics and image generation technology. Today, Evans & Sutherland designs, builds and sells advanced 3D graphics systems for modeling, visualization, simulation and virtual reality.

A publicly owned company since 1978, Evans & Sutherland common stock is traded under the NASDAQ symbol ESCC. The company is based in Salt Lake City, Utah, on a 36-acre campus in the University of Utah Research Park, and has 1,100 employees worldwide. In addition to NASA and the aerospace companies, Evans & Sutherland's products are used by most major airlines, U.S. and international armed forces, auto companies, national laboratories and pharmaceutical companies. Fiscal 1993 revenues were $142,300,000.

Evans & Sutherland has provided visual systems to aid in astronaut training since 1976. This image of a future event shows space shuttle docking with space station. (ESIG-3000.)

Evans & Sutherland offers a full range of computer visual systems for vehicle operator training, vehicle engineering, education and entertainment, as well as high-performance 3D graphics hardware and software for leading workstation vendors. The company also develops and markets computer-aided industrial design software for automobile and consumer product design.

Visual Systems for Simulation

Evans & Sutherland provides the most complete line of visual systems products available from any company for use in vehicle simulators. These systems create dynamic, high-quality, out-the-window scenes that represent the view vehicle operators would see if performing tasks under actual operating conditions. Visual

Evans & Sutherland has supplied visual systems for simulation and training applications to military and commercial customers for more than 26 years. (ESIG 3000.)

system products include EaSIEST® database modeling tools (for creating visual environment models), image generators and display systems. Evans & Sutherland visual systems are an integral part of full simulators, which incorporate a number of other components including cockpits or vehicle cabs and large hydraulic motion systems.

The ESIG line of image generators includes the midrange ESIG®-200 and the best-selling ESIG®-500, the preferred choice in airline pilot training worldwide. The ESIG®-2000, a compact, low-cost image generator that incorporates much of the technology and architecture of larger systems, has dominated the ground-warfare training market worldwide since its introduction in 1991. It is also currently being used in new commercial market activities, such as entertainment attractions. The ESIG®-3000 family consists of modular, high-performance image generators used in advanced engineering simulators and military training systems. The ESIG-3000 line of image generators has been recently introduced into the civil-aviation simulation market. The top-of-the-line ESIG®-4000 offers unparalleled scene realism for demanding mission rehearsal applications.

Evans & Sutherland has taken these proven technologies and designed the next generation of computer graphics systems, the ESIG high-density series. The series' increased processing power allows higher complexity databases for optimum simulator training.

Display systems range from simple monitors to very complex devices consisting of projectors and large domes. VistaView™ is a head-tracking projection system that provides a high-resolution picture within a pilot's immediate view. TargetView™ is a target projector that displays a fast-moving target image inside a dome. NiteView™ is a display system that simulates night-vision goggles.

Evans & Sutherland also sells DIGISTAR®, currently the world's only digital planetarium projector. In addition to projecting traditional astronomical features, DIGISTAR can display many special effects, such as star movement through time, without being restricted by mechanical motion. DIGISTAR can also be customized to allow audiences to see what it would be like to travel through the human body, or even inside a molecule or a fractal structure.

Evans & Sutherland's high-end visual systems provide sophisticated engineering problem-solving capabilities. (ESIG-2000.)

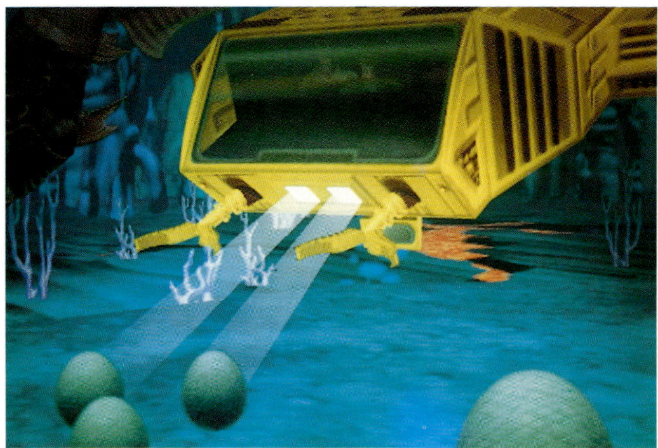

Virtual Adventures is the world's first high-capacity 3D immersive virtual reality attraction from Iwerks Entertainment and Evans & Sutherland. (ESIG-2000.)

3D Workstation Graphics

Evans & Sutherland provides 3D graphics hardware and software for leading workstation vendors. Current products include Freedom Series™ graphics accelerators and 3D application programming interfaces (APIs).

Freedom Series graphics accelerators are available directly from IBM, HP, and Sun Microsystems. The Freedom Series has the highest Graphics Performance Characterization (GPC) PLB benchmark results of any graphics system, and supports advanced graphics features like texture mapping and transparency. Freedom Series markets include scientific visualization, geoscience, industrial design, visual simulation, and high-end MCAD.

Evans & Sutherland also provides GL-based application development software tool kits for leading workstation vendors.

Design Software

Computer-aided industrial design software for automobile and consumer product design is also developed and marketed by Evans & Sutherland. The principal products are CDRS™ (Conceptual Design and Rendering Software) and 3D Paint™

CDRS is used by automotive, product and industrial designers to create, visualize and analyze product concepts and designs. The central advantage of CDRS is that product models made by these creative designers in the concept phase are usable for the actual engineering and manufacturing of the product.

3D Paint, a real-time, multi-platform digital painting software product, operates both in 2D and 3D. Using 3D Paint, designers and artists apply color, texture and artwork directly to the surfaces of 3D computer models, as if painting on a physical model with a brush. 3D Paint is fully integrated with CDRS.

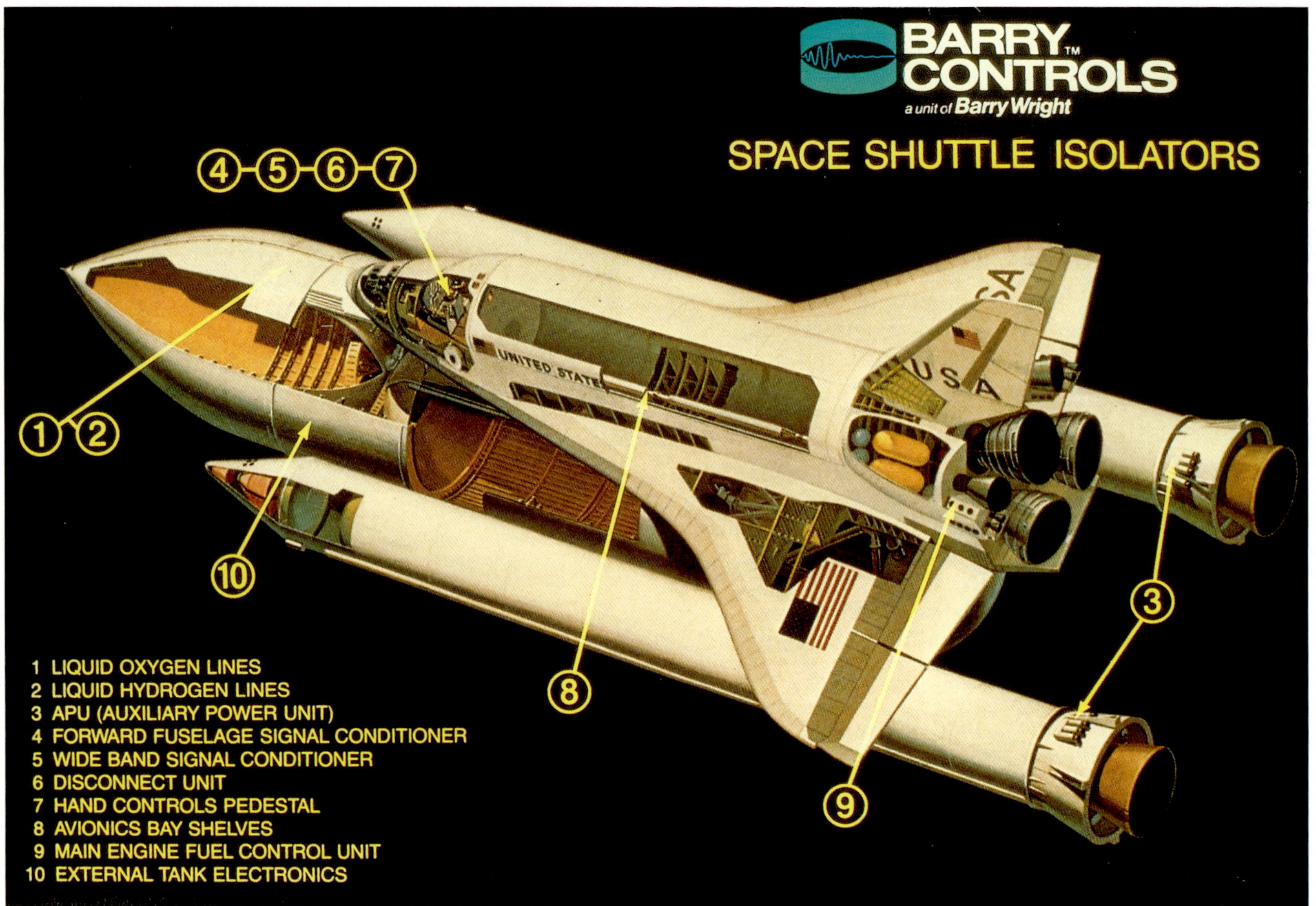

Barry Controls isolators protect the space shuttle's electronic and mechanical subsystems from the tremendous forces generated during lift-off.

Barry Controls

Each time the space shuttle blasts into orbit, Barry Controls isolation systems protect the shuttle's oxygen and fuel lines, and electronic and mechanical subsystems from the tremendous vibrations associated with the send-off.

Barry Controls is the leading supplier of innovative, cost effective solutions that reduce the effects of dynamic disturbance by isolating or attenuating shock, vibration, noise and motion in a variety of defense, industrial and aerospace applications. The company designs and manufactures standard and customized shock, vibration, and noise isolators and motion accommodation systems which provide vibration isolation for engines, electronic equipment, and navigation instruments.

Founded in 1943 to solve problems of shock and vibration for the military services, Barry Controls' first product was the cupmount, which developed from a single isolator into a comprehensive series which is still in common use today. Barry Controls continues to be a leading developer of isolation solutions for the world's defense industries. Over the last 50 years, Barry Controls has established itself as a major force in the industrial and commercial aerospace markets. Today, Barry Controls is an OEM supplier for aircraft, industrial equipment, defense equipment and consumer products. The company has more than 6,000 customers worldwide.

Barry Controls maintains its international leadership role by providing customers with "engineered solutions" that help them produce optimally performing products. State-of-the-art capabilities in advanced systems development, manufacturing processes and materials technologies have allowed Barry Controls to effectively develop new applications to meet each customer's unique needs. The company has an expert team of design engineers, chemists and dynamicists who develop customized, high-performance vibration isolation products for defense and commercial applications.

Located in Burbank, in the midst of Southern California's aerospace industry, Barry Controls Aerospace maintains a building complex containing 120,000 square feet of space devoted to the design, development and manufacture of aerospace products. Here, engineers use the latest computer-aided design techniques to perform all necessary stress and dynamic analysis to support each isolation system, and develop the resilient elements, structural materials and manu-

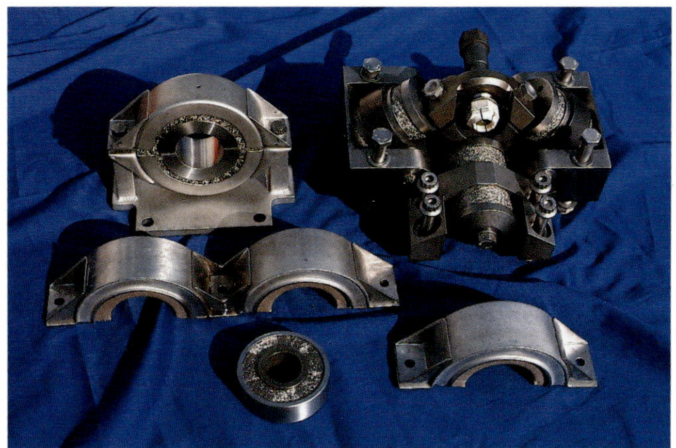

Shuttle Solid Booster Auxiliary Power Unit Isolator.

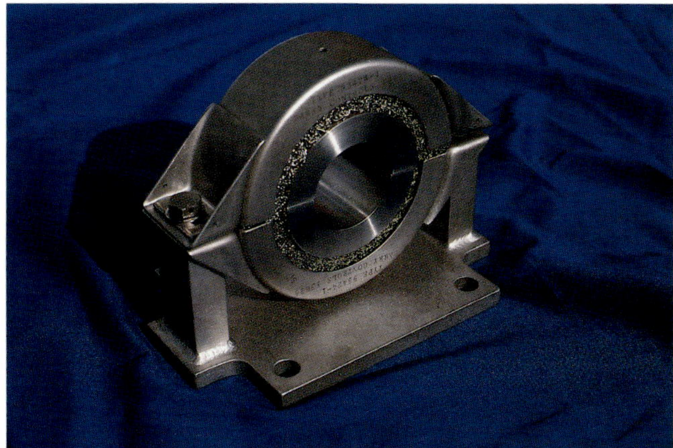

Shuttle Solid Booster Fuel Valve Isolator.

facturing processes for components that operate in such extreme conditions as space.

Barry Controls offers a wide choice of resilient elements manufactured from black rubber, Silicone, and the company's own Met-L-Flex® materials. Met-L-Flex® resilient elements are knitted stainless steel wire formed under high pressure to a shape and destiny. These elements provide uniform performance over a wide temperature range from -130° to 500° F. Special designs operate at temperatures exceeding 1000° F.

Closer to Earth, vibration, shock and noise solutions provided by Barry Controls includes aircraft auxiliary power unit isolation systems for business and general aviation aircraft and all widebody aircraft; and commercial aircraft engine vibration isolators for jet transport, all classes of turboprops, and business and general aviation aircraft throughout the world. Barry Controls is developing and manufacturing the engine vibration isolation system for the industry's newest high speed turboprop regional commuter aircraft, the Indonesian IPTN-N250, and is also providing Beech Aircraft with a newly designed and manufactured engine vibration isolation system for the regional commuter plane, the Beech 1900D. In addition, Barry Controls is enhancing its position as the world's largest supplier of aircraft cabin tuned dampers by supplying the majority of the world's turboprop commuter aircraft with a product that substantially reduces interior cabin noise levels. To meet the increasing needs of its aerospace customers and supply product at the lowest cost to a global market, Barry Controls has reorganized its aerospace operations into customer-focused business units.

Barry Controls has expanded its industrial market base with products for a variety of application areas, including automotive, trucking, railway, transportation, shipbuilding, computers, medical and laboratory instrumentation, and household appliances.

Serving the automotive industry, Barry Controls designs customized vibration isolators and mounts for the engines, cabs, radiators, mufflers, electrical components, accessories and suspension spring pivots critical to off-highway vehicles, trucks and buses.

Shuttle Single and Double External Tank Line Mounts.

Having worked with Harley Davidson for many years, Barry Controls was recently given the challenge of designing the engine mounting system for a new bike that would isolate motor vibration while retaining the classic "Harley feel." The bike appeals to a new breed of riders looking for a kinder, gentler ride without losing the traditional Harley "toughness." Barry Controls engineers modeled, developed and tested an innovative engine mounting system with an optimized geometry and material configuration that successfully met the tough design constraints as well as cost and performance expectations.

Asked to provide a high performance, long life isolation product for a shock bumper on one of Stanley Bostitch's commercial nail driver products, Barry Controls chemists developed a specially formulated elastomer material that not only met Stanley Bostitch's product life needs, but also exceeded its durability requirements.

Barry Controls is a business unit of Applied Power Inc's Engineered Solutions Group, with operations in North America, Europe, Japan and Asia Pacific, and Latin America. Barry Controls U.S. operations are located at 40 Guest Street, Brighton, Massachusetts, and 4510 Vanowen Street, Burbank, California.

Harris Corporation
Government Aerospace Systems Division

Harris Corporation's Government Aerospace Systems Division has participated in many of the major U.S. space programs since the 1950s. Harris Aerospace also supports future space exploration and eagerly anticipates a role in establishing a lunar colony and missions to Mars as the next century rapidly approaches.

A major supplier of custom aircraft and spaceborne communication and information-processing systems, Harris Aerospace has been the leader in state-of-the-art technologies for space applications since developing telemetry systems for the Telstar satellite in the early 1960s. Since then, Harris multiplex systems have provided data sampling and processing functions for Minuteman intercontinental ballistic missiles through three design generations. Harris telemetry equipment has flown on every manned American space vehicle since Alan Shepard's suborbital flight. These include Mercury, Gemini, Apollo and most recently the Space Shuttle.

Space Shuttle Atlantis

Harris' Payload Timing Buffer and Pulse Code Modulation Master Unit were on board the first shuttle mission in 1981, and these units were joined by a third device, the Payload Data Interleaver, on the orbitor's fifth mission a few years later. The buffer serves as a master clock distributor, providing signals that keep shuttle functions synchronized, and the interleaver gathers and arranges input from both onboard and remote payloads. The master unit manages all of the communication of digital status and flight dynamics information between the orbitor and Mission Control.

In addition to the shuttle systems, other Harris Aerospace's space technologies include subsystems for the Tracking and Data Relay Satellite (TDRS), NASA's spaceborne scatterometer, Defense Support Programs, the guidance sensor for NASA's Hubble Space Telescope, and communications systems for the Brilliant Eyes Satellite System and the Space Station.

NASA's Hubble Space Telescope

Space Station

The manned space station is NASA's first step in exploring the nearby solar system – the first step that is absolutely essential to all future manned space exploration. The space station will be a platform and development area for missions both to the moon and Mars. Harris' involvement with the space station is with NASA Marshall Space Center and Boeing Aerospace on Work Package 1.

International Space Station

Harris' advanced technologies will support the station modules that contain the work laboratories, the astronaut living quarters, and the logistic supply space. The system that Harris will provide is the audio and video subsystem that serves as the eyes, ears and voice of the station. The Audio and Video Distribution System (AVDS) provides audio and video communication between the crew, ground, other spacecraft and astronauts in extravehicular activity. The system will also provide entertainment for the crew, emergency communication, and audio and video monitoring of the experiments onboard the space station.

Harris' Audio/Video Distribution System

Another Harris technology concerned with the integration and communication of multiple data sources is in development – multisensor fusion technology. Fusion is the enabling technology that allows the enormous volumes of information from different sensors and data sources to be consolidated, correlated and displayed without overwhelming an operator. The Harris fusion software architecture also bridges the challenge of joining off-board intelligence data from national space assets to tactical platforms with their own on-board sensor systems that will provide real-time high resolution pictures.

The fusion software is modular and based on open system processing, so additional sensors or information streams can be readily accepted without major changes. This capability is extremely important as more and more data links are becoming operational. One real benefit of this capability is that stealthy platforms, who want to remain passive with their own sensors, can still have a wealth of intelligence and tactical information.

The fusion technology bridges into most space-based applications of information integration. These applications include the demanding tasks of the BMDO program, the ability to rapidly fuse information from space-based and ground-based sensors, the ability to neutralize ballistic missiles and critical environmental surveillance from space-based sensors.

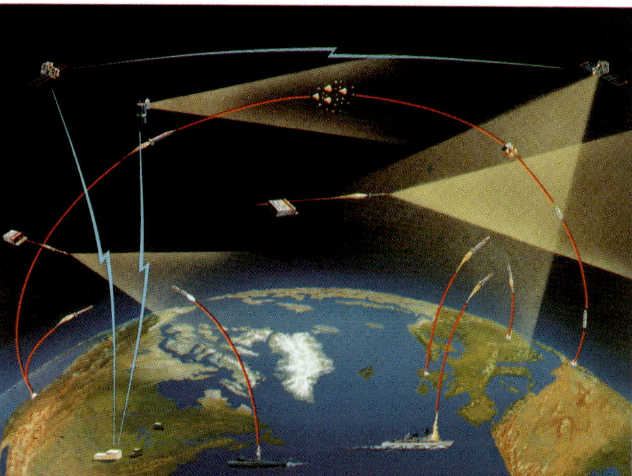

Multisensor fusion technology supporting space applications

Future Space Exploration

Future Harris systems will face stringent requirements for complex digital processing and communication systems. Reductions in system size, weight and power – along with the ability to operate in space and radiation environments – will be necessary. But, as the United States and its international partners explore space via a manned space station, a colony on the moon, and missions to Mars and beyond, Harris Government Aerospace Division will be there, providing the necessary advanced communication and information technology to support these ventures.

Harris supports future lunar-based space exploration

McDonnell Douglas Aerospace

With more than three decades of leadership in American space technology development, McDonnell Douglas is applying its innovative expertise to find new, more cost effective solutions to the 21st century challenges of humans living and working in space — and the means of transporting them there.

At the space and defense systems unit of McDonnell Douglas Aerospace in Huntington Beach, California, the scientists, engineers, technicians, and production and support people who helped write the country's success story in manned space flight continue to bridge the gap between concept and hardware to establish the world's future in space. They are working on launch vehicles, single-stage rockets, space station, and strategic defense and electronic systems.

Delta — the 'Workhorse' Rocket

Since 1960, the Delta medium-range launch vehicle — McDonnell Douglas' "workhorse" rocket — has boosted satellites into space for military, NASA and commercial users. Adapted over the years to meet more sophisticated and heavier payloads, Delta has a lifetime success rate better than 95 percent.

Delta II 7925, the latest and most powerful version of the rocket, first flew in 1990 and has lift capability to boost 4000 pounds to geosynchronous transfer orbit. It is used primarily to launch U.S. Air Force global positioning system (GPS) satellites, which form a navigation satellite system that provides "pinpoint" position and velocity information for users anywhere on the globe. McDonnell Douglas began launches for the U.S. Air Force in 1989, and a 1993 contract covers up to 36 launches through the year 2002.

A McDonnell Douglas Delta II rocket blasts into space from Cape Canaveral carrying a Global Positioning System satellite for the U.S. Air Force.

For NASA, a McDonnell Douglas Delta II has launched the Geotail and Wind satellites under the Medium Expendable Launch Vehicle Services contract. The 1991 contract calls for launch of the Polar, Radarsat, X-Ray Timing Explorer, the Near Earth Asteroid Rendezvous, Mars Pathfinder and the Advanced Composition Explorer satellites, with options for six additional launches.

Upcoming commercial launches include two Koreasat launches for Korea Telecom, planned for 1995, and the Skynet 4D for the United Kingdom Ministry of Defence, scheduled for 1997. Beginning in 1996, McDonnell Douglas' Delta IIs will launch 40 satellites for Motorola's Satellite Communications Division to form the major portion of the global wireless IRIDIUM® telecommunications network.

Single-Stage Rocket Development

Aiding NASA's efforts to develop reliable, low-cost space transportation, McDonnell Douglas Aerospace continues integration of advanced technology components into and testing of the Delta Clipper-Experimental (DC-X), a one-third scale version of a single-stage, reusable rocket that takes off and lands vertically, but is maintained and operated in the same manner as an airplane.

Developed under contract to the Ballistic Missile Defense Organization, the DC-X stands 40 feet tall, is approximately 13.5 feet across its base heat shield, and weighs 41,630 pounds at takeoff. The vehicle is fueled with liquid oxygen and liquid hydrogen and is supervised from the ground by a crew of three in the Flight Operations Control Center. There the flight manager can adjust computer commands to accommodate changes in flight conditions and can direct the vehicle to terminate its flight and land if conditions dictate. The DC-X was successfully ground and flight tested in 1993 and 1994.

McDonnell Douglas is teamed with Boeing to compete for a follow-on program and hopes eventually to build an operational single-stage-to-orbit rocket able to fly orbital or suborbital missions, manned or unmanned.

Space Station — The Next Step

Since the first manned Mercury flight, McDonnell Douglas has contributed to every aspect of America's manned space program: Gemini, Apollo, Skylab, Space Shuttle, and now Space Station. The company continues work on various space station design ele-

In a 1993 vehicle hover test at White Sands Missile Range, New Mexico, the DC-X (left to right) lifted off vertically, hovered at about 150 feet, then moved horizontally to land in a vertical position on a different pad. (Photos by George Baird, U.S. Army White Sands Missile Range, N.M.)

ments, such as truss structure, resource nodes, airlocks, and mobile transporter, as well as distributed systems, such as guidance, navigation and control, data management, thermal control, communication, and tracking.

Space station will provide a permanent work base for humans studying the earth's resources, developing new manufacturing and processing techniques, examining the origin and evolution of the universe, and performing basic science and technology research.

Defense and Electronic Systems

McDonnell Douglas Aerospace's defense and electronic systems work encompasses programs in laser systems, surveillance and detection, communication systems and strategic defense. The unit's Mast Mounted Sight (MMS), a surveillance and target acquisition system, can be mounted on aircraft, ships or fixed sites and provides a 360-degree view for reconnaissance and attack missions. To complement MMS, the company developed a smaller, advanced system known as NightHawk, which has both military and commercial uses.

The defense and electronic systems group also designs and manufactures avionics and ground support electronics for military and civil aircraft, missiles, and the Delta launch vehicle, and produces semiconductor lasers for use in military and industrial laser products.

Transport Aircraft

McDonnell Douglas Aerospace is also comprised of the military transport aircraft and advanced transport aircraft design units.

The primary focus of the transport aircraft unit is the U.S. Air Force C-17 Globemaster III. Designed to fulfill airlift needs well into the next century, the C-17 will carry large combat equipment loads and troops or humanitarian aid anywhere in the world. A unique feature of the aircraft is its ability to land in small, austere airfields, minimizing the need to move troops and cargo over land.

The advanced transport aircraft development unit is developing technology that will be incorporated in the design and manufacture of current and future civilian and military transport aircraft, including supersonic civilian transport aircraft.

The C-17 Globemaster III transport, equipped with McDonnell Douglas' airlift defensive system which protects the aircraft against surface-to-air missiles with infrared seekers, undergoes testing at Eglin Air Force Base in Florida.

Applied Solar Energy Corporation

Electrical power for all of the world's satellite systems is the result of technology developed by Applied Solar Energy Corporation, the most experienced space photovoltaic cell manufacturer in the world.

Founded in 1955, Applied Solar Energy Corporation has been in the forefront of space solar cell technology for the past 40 years. Since the 1958 flight of Vanguard I, Applied Solar's solar cell technology has been used to power an unmatched line-up of NASA scientific, military defense, and commercial communication satellites. The company is a major contributor to the United States' position as world leader in space satellite technology.

Clementine Interstage Array.

Space Solar Products

Applied Solar supplies photovoltaic solar cells, cell assemblies, panels and arrays for space electrical power. Photovoltaic solar cells generate power by capturing and converting the sun's radiating photon energy into electricity. The company's space solar cell technology includes expertise in silicon, gallium arsenide and cascade multi-junction devices.

Advanced Technology

Applied Solar has been chosen by the United States to develop and transition into production, three successive generations of technology for solar cells including the state-of-the-art dual-junction, cascade cells. These cells, with an average efficiency of 22 percent, offer unprecedented system power for military, scientific, and commercial satellites. Applied Solar has also been selected to develop a triple-junction cell which promises to yield even higher efficiencies and will become the standard technology for future satellites.

A critical component of the development of succeeding generations of higher efficiency solar cells is the unique ability to transition the cell designs into high volume production. Applied Solar has repeatedly demonstrated this important capability. The GaAs/Ge cell technology, which, along with silicon, has been the standard power source prior to the advent of cascade cells, was developed in the 1980s at Applied Solar. Since then, Applied Solar has manufactured more than one million GaAs/Ge cells and delivered 98 percent of all GaAs/Ge cells in the world. The dual-junction cascade cells are now into production with the triple-junction cells to follow.

The continued confidence of the U.S. government and world-wide satellite manufacturers in Applied Solar's ability to develop and produce the next generation solar cells is a direct result of Applied Solar's past successes and the company's commitment to maintaining its leadership position in photovoltaic technology.

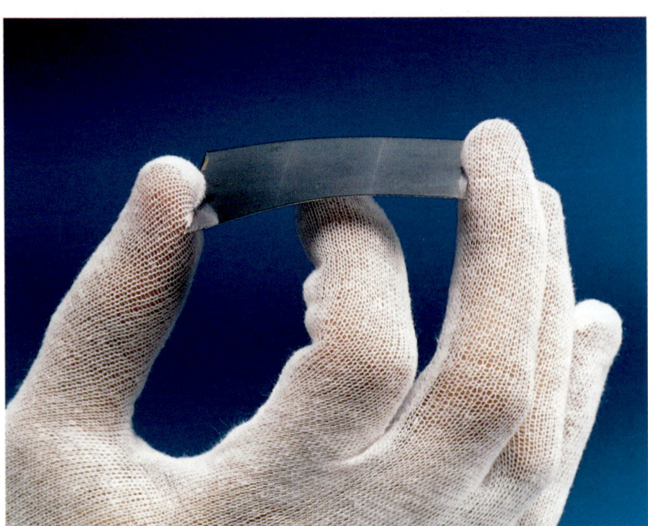

Large Area GaAs/Ge Solar Cell.

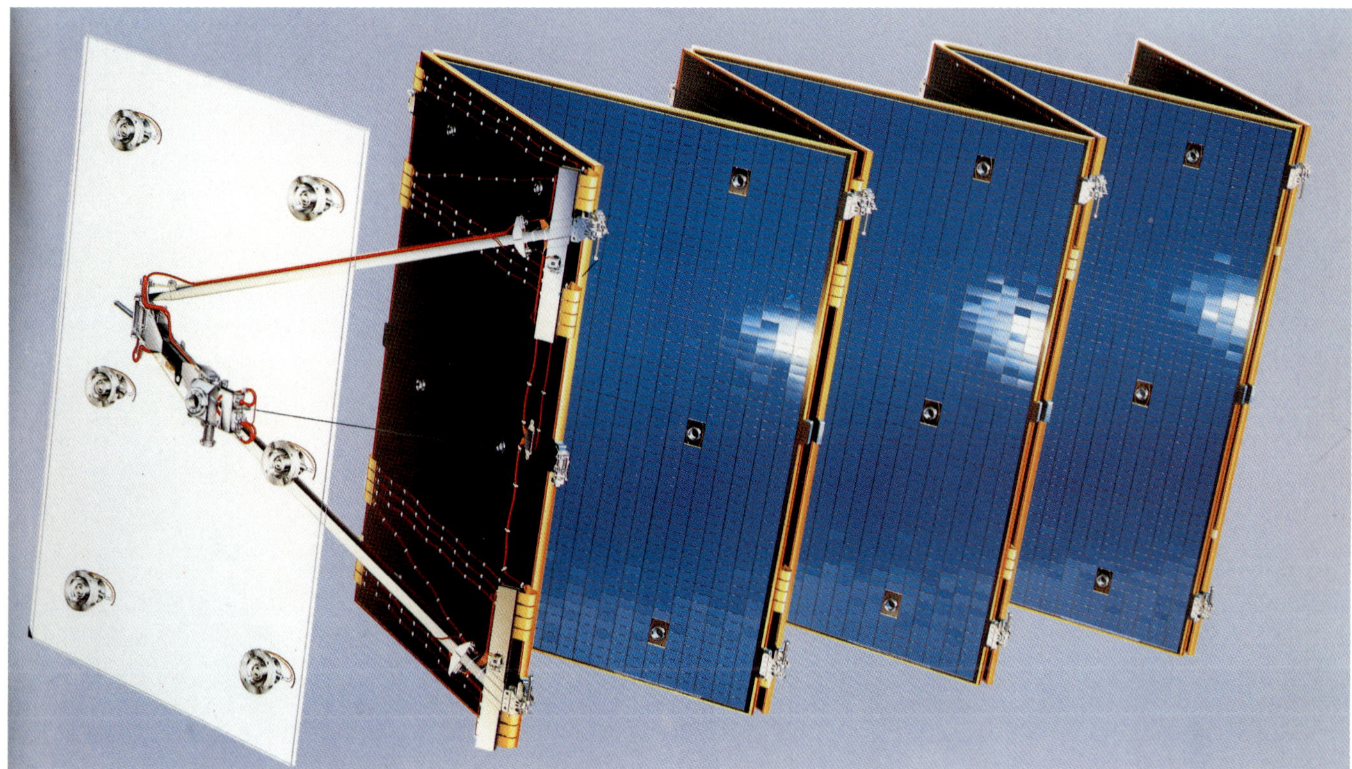

High-power Multi-panel Solar Array.

Gallium Arsenide and Silicon Solar Cell Assemblies

Applied Solar has manufactured silicon solar cells since the 1950s and gallium arsenide/germanium cells since 1986. Both the silicon and gallium arsenide technologies exhibited steady increases in the energy conversion efficiencies as the technology matured.

In the case of gallium arsenide, Applied Solar routinely offers average lot efficiencies of 18.5 percent and is capable of producing large quantity batches in excess of 19 percent efficiency should a mission require such power levels.

Applied Solar manufactures both gallium and silicon cell assemblies in various sizes and thicknesses.

The gallium cells can be customized in sizes up to 100 square centimeters as a program may require, or can be delivered in standard production sizes: 6.5 cm x 5.5 cm; 4.0 cm x 4.0 cm; 4.4 cm x 4.0 cm; 2 cm x 4 cm, etc., which allows for cost advantages of economies of scale and lower non-recurring efforts. These cells are processed in thicknesses from 3.5 mils to 8 mils.

Silicon cells are available up to 64 square centimeters with similar standard sizes to gallium arsenide. The thickness ranges from 2 mils to 10 mils. The average efficiency ranges from 12 per cent to 17 percent AMO.

Applied Solar provides solar cells in a variety of assembly configurations. The most common of these is the cell-interconnect-coverglass (CIC) configuration, which provides a complete assembly of each specified component ready for stringing and laydown.

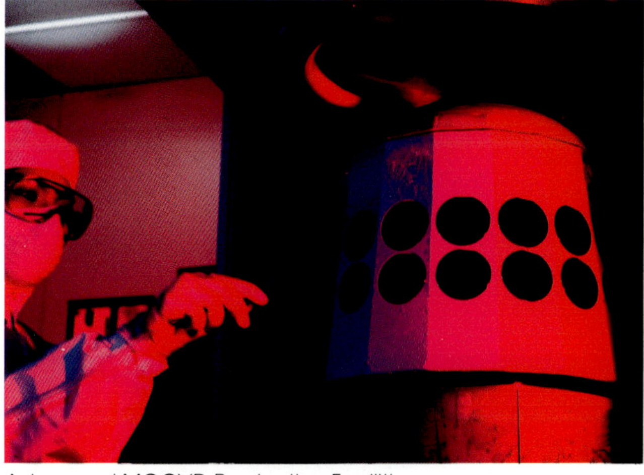

Advanced MOCVD Production Facilities.

Gallium Arsenide and Silicon Solar Panels

Applied Solar has completed construction of a new 7,500 square-foot medium and large solar panel production facility. Gallium and silicon panels are assembled in this area, which can accommodate panels up to 10 square feet.

With a high level of panel laydown expertise, Applied Solar offers a complete solar panel product line to the worldwide satellite manufacturers. With panel design, test and assembly capabilities complimenting the largest space solar cell assembly capacity in the world, Applied Solar provides a vertically integrated solution to the satellite power requirements.

Applied Solar has approved welding and soldering processes for both gallium arsenide and silicon cells to

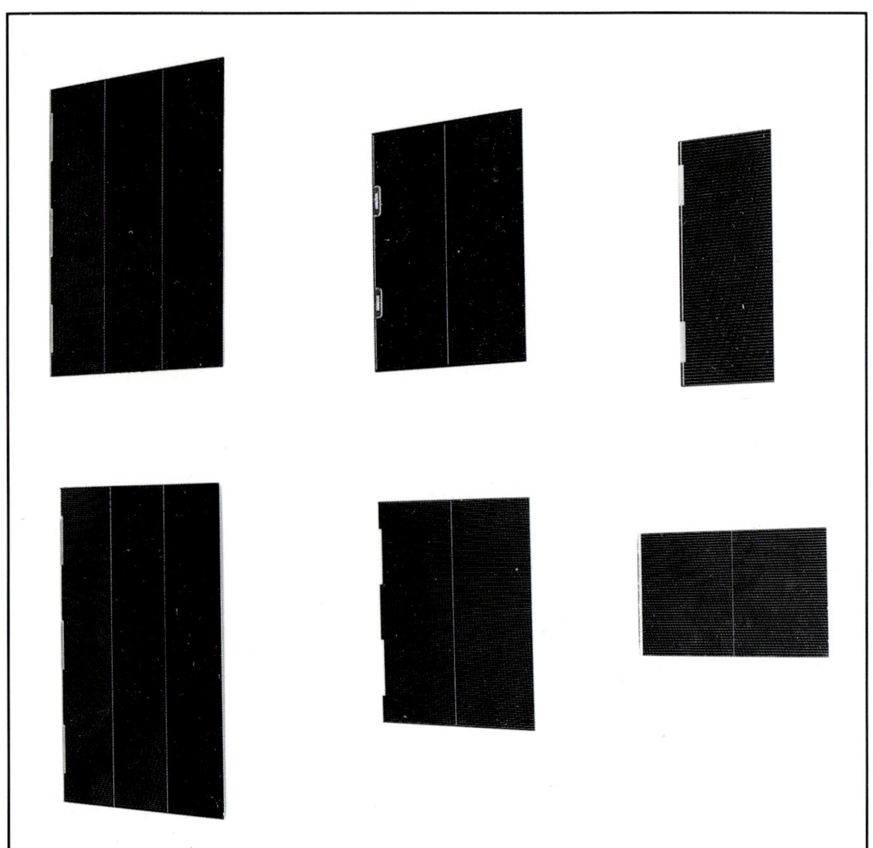

GaAs Cells from World's Largest Production Facility.

provide design flexibility to its customers. They provide detailed power analysis summaries as well as design to cost optimization of cell sizes, coverglass, interconnects, etc.

Recent panel build projects combine a mixture of commercial, NASA and military satellites, including LANDSAT 7, GPS, CLEMENTINE, FORTE, ORBCOMM, GFO, and Mars Pathfinder, among others.

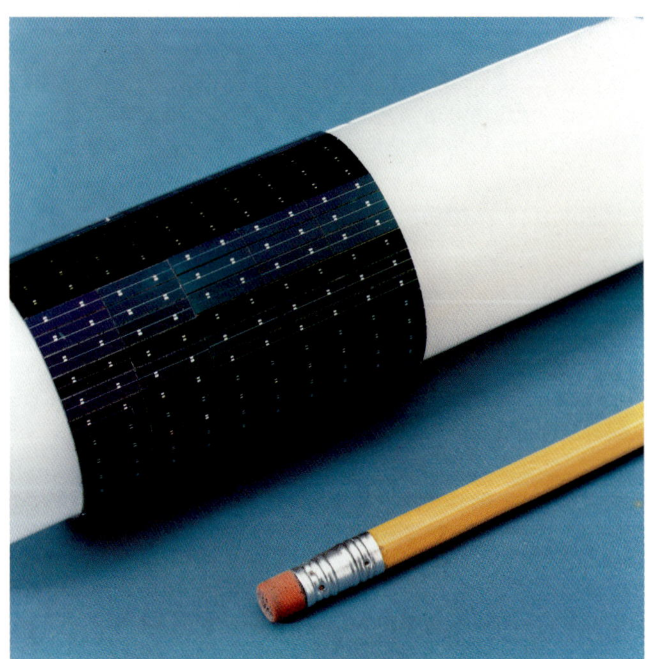

Advanced Micro Array Technologies.

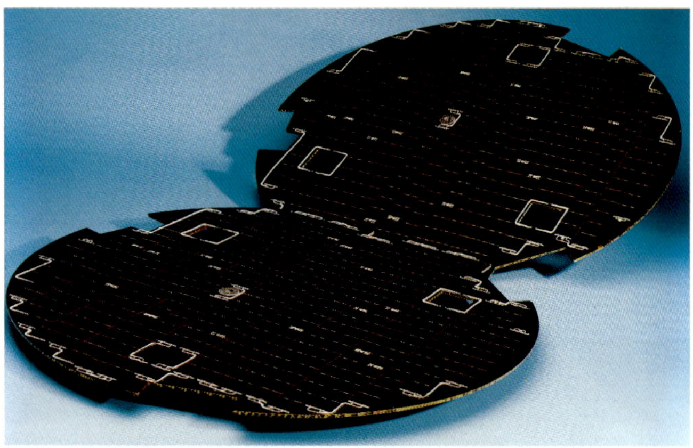

Production Silicon Arrays.

Facilities

Applied Solar is located in the City of Industry, California, in a modern 150,000 square-foot facility.

The gallium arsenide and silicon cell production lines utilize state-of-the-art semiconductor manufacturing and test equipment. Applied Solar is unmatched in the world with capacity in excess of 375 KW per year for gallium and silicon cells. Equivalent capacity for cascade cells is being installed at the time of this publication.

Currently, hundreds of thousands of space solar cells, which assure reliable, failure-free power in outer space, are produced annually.

The panel assembly facilities incorporate the latest in welding, soldering, fixturing and test equipment. This facilitates a high volume throughput for small, medium and large panel assemblies.

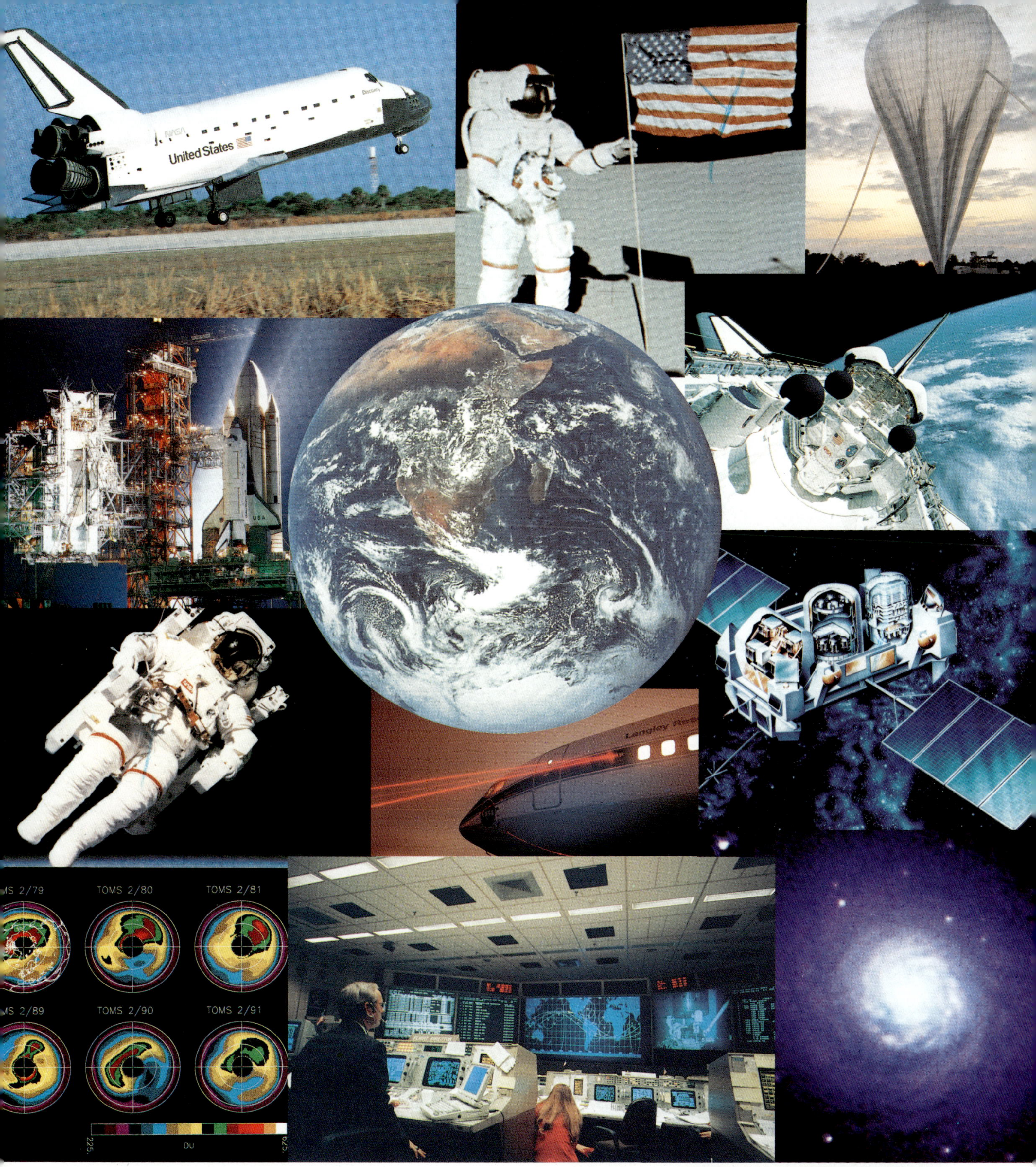

Supporting the NASA Mission

digital™

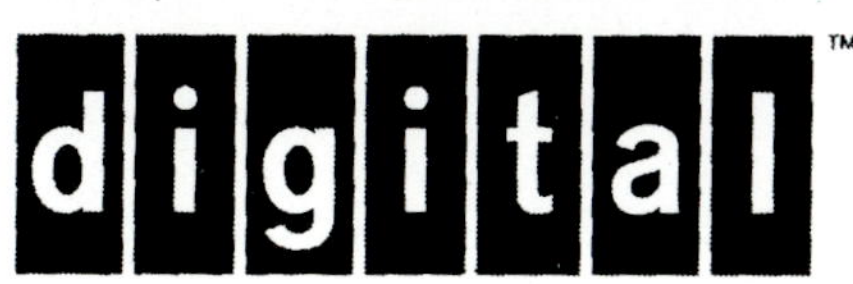

HUGHES TRAINING, INC.
Link Division

Humans living and working in space environments in the 21st Century will be trained for their tasks on systems now under development at Link. Link is NASA's prime training systems contractor for manned spaceflight.

For more than 60 years, the Link[R] name has represented the standard of excellence in simulation technology and training services for the U.S. military, international forces, NASA, government and business customers.

For the past 33 years, Link has been simulating the sights, sounds and feel of space — and the machines that carry humans there. Link simulators have trained astronauts and flight controllers, beginning with the Gemini program through Apollo, Skylab, Space Shuttle, Spacelab, and Space Station.

The company first employed simulation techniques in 1962 to help train astronauts for pioneering Gemini flights. Apollo astronauts trained for their moon missions in Link command module and lunar module simulators. These simulators played a vital role in the rescue of the Apollo 13 astronauts after an explosion in the service module severely depleted Apollo's power systems.

Upon landing Space Shuttle *Columbia* in the California desert after its first flight in 1981, astronaut John Young, spacecraft commander, praised Link for the realistic training its simulators had provided. "We had a thousand landings in the simulator, and the vehicle flies just like the simulator," he said.

Astronaut John W. Young, seated at the flight deck of space shuttle *Columbia*, praised Link for the realistic training its shuttle simulators provided. "The vehicle flies just like the simulator," he said upon landing Columbia after its first flight in 1981.

Link Simulation Technology Makes the Impossible Possible

Training for spaceflight with its attendant simulations is unique in that the total space system and environment must be created and trained in long before the first flight. There is no luxury of "field tests," "test pilots," or "test flights." Each is for real and final.

With the capability to control the environment, quickly collect multiple data on alternative designs and user reactions, and test scenarios not ordinarily available, simulation has proven to be indispensable for astronaut training and a viable and cost-effective supplement to field testing of prototype real-world equipment. A hallmark of Link's spacecraft simulations has been the use of the spacecraft's flight software in the simulator.

Achieving full-fidelity realism often requires replicating visual, motion and sound effects which serve as important cues in operating equipment or evaluating a situation. Link researchers have invented many of the cueing technologies in common use today, and continue to refine them and develop new approaches to satisfy different customer requirements. Among those technologies are helmet mounted displays, control loading, motion systems, G-seat/G-suit, and aural cueing.

The fidelity of the visual cues provided by the Link-developed ClearVue™ helmet mounted display is only limited by the state-of-the-art in the image generation equipment. Simulated visual scenes are presented to the person wearing the helmet through little television screens in front of each eye. The ClearVue™ technology allows training to be conducted with computer-generated images and real spacecraft or spacesuit hardware.

Link developed TrueSound™, a virtually noise-free digital audio simulation for aural cue and audio interconnection requirements. TrueSound has been applied to NASA's Shuttle Mission Simulator, Lockheed-Martin's Advanced Tactical Fighter research, classified naval programs, air traffic control and air defense training.

It is commonly known that the industry-standard six-degree-of-freedom synergistic motion system is a Link invention. The newest version features state-of-the-art hydraulics and accommodates heavier payloads. The company also makes three-degree-of-freedom systems for flight training devices, railroad simulators and other needs.

What is not commonly known is the space environment demanded a seventh degree of freedom for shuttle launches. Link provided this innovative, one-of-a-kind device.

Right: Link's Houston Operations is currently developing the Space Station Training Facility in the same building that houses the Link-developed Shuttle Mission Simulator that trained John Young. In addition to Space Station astronaut training, it will validate Space Station design concepts.

Manned Spacecraft Training Systems

The Shuttle Mission Simulator has a motion-based crew station mounted on the seven-degree-of-freedom motion system and two fixed-base crew stations to provide full mission training for shuttle crews. Included are launch, orbital insertion, orbital operations, reentry and landing, as well as integrated training with the Mission Control Center. The Shuttle Mission Simulator is regularly upgraded to remain current with the shuttle program's expanding technology.

The Spacelab Simulator trains astronauts to operate and perform experiments in the European Space Agency-developed Spacelab module during shuttle orbital operations. The Spacelab Simulator is co-located and networked with the Shuttle Mission Simulator to provide full mission training for the highly complex Spacelab missions.

Past programs such as Apollo were essentially stand-alone, single-use, one-at-a-time programs. Training for these programs reflected this independence.

In the coming era of manned spaceflight, there will be multiple programs in various stages of maturity. The space vehicles will have international origins, have multiple uses, will operate in conjunction with one another creating numerous and complex interdependencies between programs, and, because of the long life of the programs, will represent an ever-increasing, cumulative development, operational, and support workload.

Therefore, the Shuttle and Space Station simulators are multi-program, interdependent and habitation-operations oriented and built on a training infrastructure that is both low cost and open to technical innovation.

Link is developing solutions to the training problems presented by the assembly and operation of the international Space Station, and is evolving the NASA training system infrastructure to meet future space exploration.

The resulting Space Station Training Facility, under development by Link's Houston Operations in the same building that houses the Link-developed Shuttle Mission Simulator that trained John Young, is the focus of Space Station astronaut training and also serves to validate Space Station design concepts.

The Space Station Training Facility has a suite of training devices, from part task trainers to a full task trainer also co-located and networked with the Shuttle Mission Simulator.

A good amount of the work on Space Station will be shared by humans and machines. A crew member, for example, might sit inside one of the modules wearing a harness and a helmet mounted display like ClearVue. Outside, a robot would inspect the station's solar panels for damage. As the robot moves methodically from one end of the solar panel to another, scanning the area with television cameras, the crew member sees through the helmet mounted display what the robot sees. If the crew member moves his or her arm inside the harness, the robot's arms would move, so the crew member becomes a teleoperator, doing work in space without the trouble and risk of having to go outside. The virtual reality technology within the Space Station Training Facility allows the crew member to train for such operations before arriving on orbit.

For more than six decades, Link has applied innovative technologies to satisfy the training and productivity needs of military, government and commercial customers worldwide. Among Link's sophisticated training systems — in addition to NASA's Shuttle Mission Simulator and the developing Space Station simulators — are the F-117A stealth fighter/bomber total training system, the B-2 aircrew training devices, the C-130 ATS (the world's largest contractor aircrew training system), the AH-64 Apache combat mission simulator (only training winner of the Daedalian Award for excellence), and the F-16 multi-national weapon systems trainer program.

On the opposite end of the training scale, Link is pioneering compact and transportable training solutions using commercial off-the-shelf equipment for desktop and classroom environments. These include multimedia computer-based training; modular, reconfigurable, low-cost flight training devices for airlines, commuter, general aviation, and military aircraft; and travel-saving media conferencing.

From classroom to outer space, Link's focus is on training, simulation, productivity and safety.

USBI Co.

From aft skirt to nose cap, the integration of the 35,000 components that make up the Solid Rocket Boosters (SRBs) which put space shuttle into orbit is the task of USBI Co., a wholly-owned subsidiary of United Technologies Corporation.

As a prime contractor to NASA, USBI also is responsible for the design, acquisition, assembly, checkout, refurbishment, recertification and flight system performance evaluation of the nonmotor segments of the SRBs. After every shuttle flight, USBI checks and refurbishes 5,000 parts for reuse.

USBI began operations in 1977 as United Space Boosters, Inc. when it won the NASA prime contract for assembly and refurbishment of the non-motor components of the space shuttle SRB. After the *Challenger* accident, NASA assigned USBI the full design responsibility for the SRB, as well as the technical integration of the SRB, including the motor segments.

To accomplish these tasks, USBI designed and operates NASA's 44-acre SRB Assembly and Refurbishment Facility at Kennedy Space Center. There the company processes booster components in a time-saving, cost-effective method. The work involves precision robotic application, removal and reapplication of multilayered protective coatings and insulation, and testing, servicing and installation of electrical, mechanical and hydraulic systems.

USBI designed, developed and installed the software and hardware for the shuttle's Automated Booster Assembly Checkout System (ABACS), which ascertains that the SRBs' functions are working properly. ABACS checks out the SRB's aft assembly, forward assembly, the guidance system gyros, range safety system and the thrust vector control system.

In addition, USBI developed a wide variety of specialized ground support equipment, including four large gantry robots for application and removal of the thermal protection system. A 30,000 gallon, aboveground washer, which circulates 1,300 gallons of water per minute, aids USBI in cleaning the parachutes which slow the SRBs to their 50 mph ocean splashdown. After drying, the 150 miles of fabric are examined and repaired for reuse.

Other USBI developments include new methods to reduce corrosion and effective support systems to manage the 35,000 components of the SRB. Each carries a USBI part number for tracking purposes.

The Dragon System

USBI's extensive expertise in design analysis, testing and flight evaluation of parachutes and gliding wings, gained from SRB recovery experience, has culminated in the development of the Deployable Ram Air Glider with Onboard Navigation, or Dragon System. The Dragon System has glide ratios up to 10:1, wind penetration capability, precision delivery, and large range distribution.

For Department of Defense purposes, this technology, coupled with an integrated guidance, navigation and control system, and linked to the global positioning system, can provide the delivery capability required to ensure accurate and flexible military support to early entry forces and war fighting sustainment. The Dragon System also has potential commercial application as unmanned aerial vehicles, in disaster relief, to provide remote access and for recreational purposes.

The Dragon System design consists of an internal rigid structural frame, totally enclosed in a ram-air inflated, double-surface sail. The frame includes mechanisms to facilitate folding of the wing for packaging.

Other Technology Transfers

Through independent research and development, USBI is pursuing new and improved applications in the areas of materials, processes, analysis technology, and system/subsystem development. Current studies are expected to lead to advanced application of automation and robotics, including new materials and processes for thermal protection, automatic testing and health monitoring, in-situ diagnostics, fiber optics and automated assembly and refurbishment methods.

About USBI Co.

USBI employs 1200 aerospace and high-tech specialists, including more than 450 professional personnel with an average of 14 years experience in various disciplines from aerospace engineering to thermodynamics.

USBI's administrative and engineering headquarters are located in four buildings of the United Technologies Corporation's 40-acre site in the Cummings Research Park, Huntsville, Alabama. USBI also operates extensive materials, chemistry and automation/robotics laboratories in approximately 175,000 square feet of modern facilities there. Other facilities include the Assembly and Refurbishment Facility at Kennedy Space Center and an Automatic Test Equipment Development Center in Slidell, Louisiana.

USBI is a wholly owned subsidiary of United Technologies Corporation, one of America's largest multinational companies. United Technologies Corporation employs more than 168,000 people around the world. Corporate headquarters are in Hartford, Connecticut.

Barrios Technology

Barrios engineers are leading the way to ensuring the United States will achieve a permanent presence in space. As the leaders in the planning and training for future operations on the Space Station, Barrios engineers are also laying the foundation for their next mission — monitoring and controlling the daily operations of the Space Station.

The people at Barrios Technology, Inc. work hard on Earth to make work in space simpler and more efficient. The company offers a complete line of products and services that efficiently utilize personnel and computer resources in the conduct of effective civil and defense space operations.

Founded in 1980 to provide mission planning and flight design services for NASA's Space Shuttle Program, Barrios has developed the knowledge and expertise necessary to analyze the entire range of tasks required to support both manned and unmanned space operations. These tasks are then designed and implemented to minimize development and recurring costs while maintaining product quality and mission effectiveness.

Barrios Space Operations expertise includes concepts and requirements definition, mission planning and analysis, real-time operations, astronaut and flight controller training, software tool development, systems engineering and integration, payload integration, and configuration management.

Barrios offers enhancements to existing software, hardware, and networks, and develops information management systems. The company is expert at converting and reengineering existing software to enable reuse on new space initiatives. A commercial division provides computer maintenance and repair services.

Barrios personnel ensure that the instrumentation used for space applications functions properly. Barrios technicians test, calibrate, and repair electrical, electronic, physical, and mechanical equipment.

Training, a cornerstone at Barrios since its founding, is provided at the Barrios Training Center, a fully equipped computer laboratory located at Barrios' corporate office in Houston, Texas. The Barrios Training Center offers a full range of UNIX courses and is also an authorized training center of Sun Microsystems and Frame Technology. NASA, the Department of Defense and other government agencies, as well as the commercial sector, take advantage of Barrios' quality training services.

Barrios' expertise has benefited a multitude of space programs: Space Station, Space Shuttle, Assured Crew Return Vehicle (ACRV), New Initiatives (Lunar/Mars), Shuttle Pallet Satellite (SPAS), Low Power Atmospheric Compensation Experiment, numerous Air Force satellite programs, and Clementine.

Barrios, a woman-owned small business, is committed to providing high quality, cost effective services to customers, as well as a rewarding work environment for employees.

The real-time monitoring and control of the SPAS II satellite by a team of Barrios engineers is just one example of the industry-leading work that has gained Barrios Technology national recognition in the field of spacecraft operations and flight planning. (Art courtesy of Mark Pestana.)

Barrios technicians are responsible for the maintenance, calibration, and repair of thousands of pieces of electronic instrumentation that are critical to the daily success of the operations at the Johnson Space Center.

Silicon Graphics, Inc.

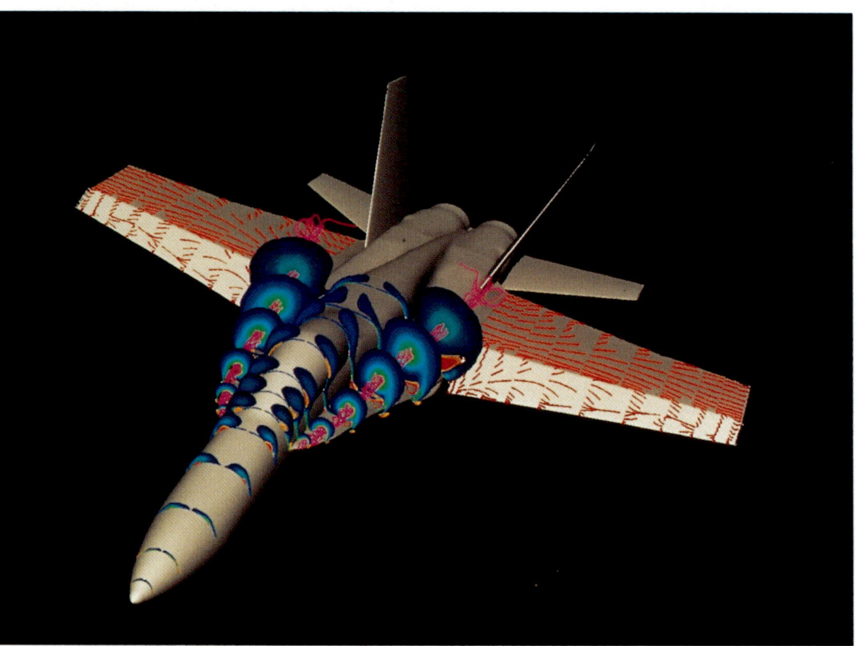

A computational model of an F-18 created with a Silicon Graphics visual computing system displays particle traces, oil flow, and helicity contours.

NASA's association with Silicon Graphics, Inc. began in 1980, when the company researched and developed the ability to visualize computational fluid dynamics data for NASA's Ames Research Center. The Interactive Raster Imaging System (IRIS®) developed for NASA was the genesis of visual computing. Based in Mountain View, California, Silicon Graphics has since become the world's leading supplier of high-performance visual computing systems. Today, Silicon Graphics works closely with NASA centers nationwide on a variety of innovative projects.

At NASA Langley Research Center, Silicon Graphics® technology serves NASA's structural analysis and mechanical engineering needs. A CHALLENGE™ L database server connected to workstations—including over 60 Indigo® and Indigo²™ desktop graphics systems—saves valuable time and expense by visually simulating, testing, modifying, and validating the design of aircraft parts for flight experiments.

Many NASA projects use visual flight simulation technology for everything from 747s to Blackhawk helicopters and fighter jets. At Langley, two multi-headed Onyx™ graphics supercomputers with RealityEngine²™ graphics subsystems create heads-down displays driving real-time flight simulations. Indigo² Extreme™ Graphics systems determine the most efficient cockpit layout. 18 displays permit the optimum arrangement of simulated readouts that look and function just as they would in a real aircraft.

NASA's Huntsville Operations Support Center (HOSC) in Alabama is the hub for the collection of information gathered on space missions. When the Space Station is completed, HOSC will gather data sent continuously from the station. HOSC turned to Silicon Graphics for the development platform and the implementation of its expanded data storage system. Information from the station as well as the space shuttle and Earth Observing System (EOS) satellites, which monitor weather and climate patterns, will be archived in CHALLENGE L and XL database servers and monitored in near-real time on up to 300 Indy™ graphics workstations. Visualized atmospheric data will be sent for analysis to such HOSC clients as the Global Hydrology and Climate Center.

The leadership of NASA and Silicon Graphics in computational fluid dynamics recently culminated in the Numerical Aerodynamic Simulation (NAS) Facility at NASA Ames. NAS enables NASA to develop system software for Computational AeroScience (CAS) projects focused on designing future generations of aerospace vehicles. The Facility selected the POWERCHALLENGEarray™ distributed parallel processing system as part of a system software testbed for distributed computing. POWERCHALLENGEarray combines 16 POWER CHALLENGE™ L supercomputers, each incorporating the world's fastest supercomputing microprocessor, MIPS® RISC R8000™, connected in a high-speed network, and will be used for CFD and related research.

Remote Access Wind Tunnel (RAWT) tests conducted by the Design Cycle Technologies Branch at Ames permit aircraft manufacturers to monitor design experiments in real time from their offices on Indy workstations, rather than sending teams to observe testing on site. Silicon Graphics InPerson™ collaboration and conferencing software allows test participants to appear on screen via camera-equipped monitors and interactively annotate images. This technology promises to strengthen partnerships among NASA and the business and scientific communities.

Silicon Graphics is proud that its pioneering technology assists NASA's exploration of the frontiers of science, and remains inspired to help NASA visualize the future.

Eaton at Home in Space

The next quarter century of achievements in space will see Eaton Corporation building upon the extensive earlier contributions of Consolidated Controls Corporation (CCC).

Since the acquisition of the two divisions of CCC by Eaton in 1986, Eaton has encouraged both the Bethel Sensors and El Segundo fluid, motion and cockpit control facilities to expand their roles as suppliers of components for space applications, continuing their proud tradition into the next exciting quarter century of space exploration.

These Eaton businesses provide unique high technology products to the major markets of energy, aerospace and defense. Eaton's business concentration is primarily the design and manufacture of sophisticated control systems and their related components and accessories. This broad product line finds extensive use in aircraft, spacecraft, industrial and power generation applications.

Founded in 1947 as the Aircraft Products division of Manning, Maxwell & Moore, it was acquired by the Condec Corporation in 1958. It has grown and diversified, priding itself on its contributions to NASA and the advancements in space exploration. Eaton components have flown on the spacecraft of all NASA programs including Gemini, Apollo, Mariner, Viking, Saturn, the Space Shuttle Transportation System (STS) and many, many more. Each shuttle launch currently utilizes more than 270 Eaton produced fluid and gas controls, transducers and ordnance devices.

Eaton's fluid, motion and cockpit controls business, located in El Segundo, California, was established in 1958. A few aircraft controls made up the entire product line. Success in the aircraft components field, coupled with an innovative engineering staff, led the company to its first endeavors at space-related fluid controls. Before long, the company was well known as an aerospace component supplier. Repeat sales and acquisition of new product lines resulted in substantial company growth.

By 1970, the El Segundo business produced pressure regulators, helium latching valves and propellant isolation valves had flown on Mercury, Gemini, Saturn, Delta/Thor, Mariner, Viking Orbiter, Lunar Orbiter, Apollo and Skylab. Further acquisition of products and expansion of the company's space related clean rooms and test facilities, together with the company's reputation for dependable products, resulted in Eaton being selected to supply more than 100 sophisticated fluid control and ordnance devices that perform critical functions throughout every phase of each space shuttle mission. Prior to lift-off, Eaton ignition safe and arm devices initiate the solid rocket boosters. Cryogenic ball valves, pressurant valves, helium regulators and helium relief valves perform in the orbiter and external tank during main engine burn. Cryogenic disconnects close orbiter and external tank side poppets prior to external tank jettison. Throughout the shuttle flight, high and low pressure solenoid valves, bi-directional solenoid valves, isolation valves and dual helium regulators perform vital functions in the orbital maneuvering system, reaction control system and auxiliary power units.

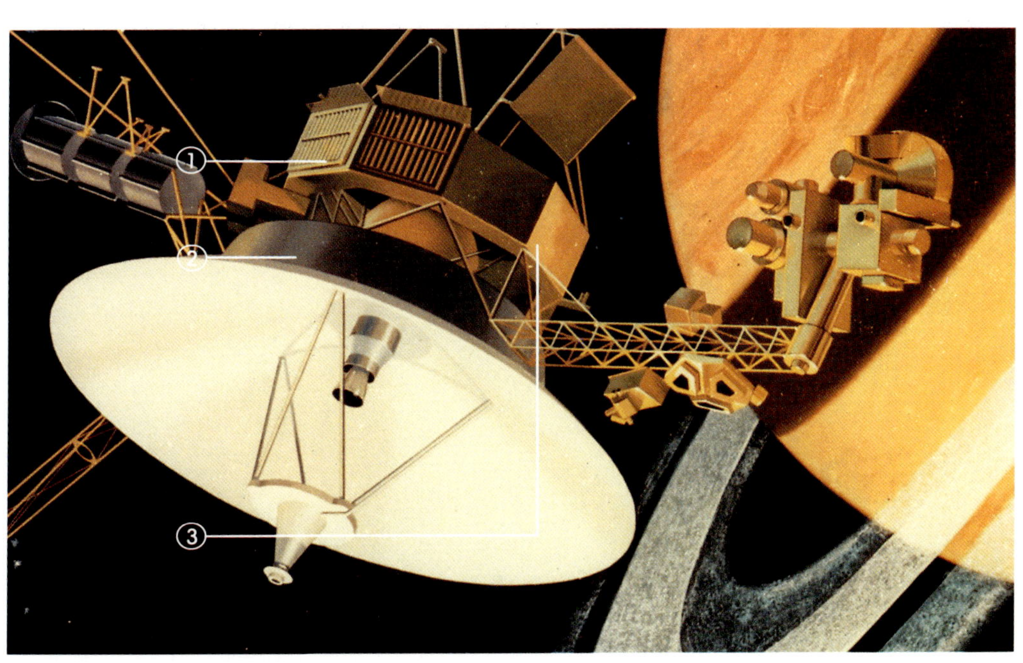

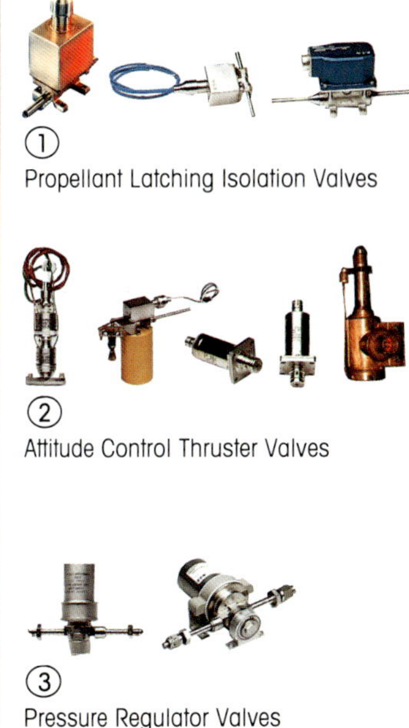

① Propellant Latching Isolation Valves

② Attitude Control Thruster Valves

③ Pressure Regulator Valves

① ②
External Tank GH_2 / GH_2 Vent / Relief Valves

③
Hydrazine Fuel Isolation Valves

④
Wing Ascent & Descent Doors

⑤
Environmental Control Latching Isolation Shut-Off Valve

⑥ ⑦
RCS Hi & Lo Pressure Solenoid Control Valves

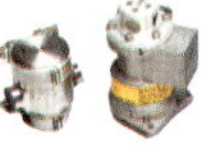

⑧ ⑨
750 PSI Pressurization Regulator / Relief Valve

⑩
Bi-Directional O_2H_2 Fuel Cell Solenoid Valves

⑪
$1^1/_2$" & 2" POGO Cryogenic Ball Shut-Off Valves

⑫
2" & 4" External Tank Disconnect Valves

⑬
20 PSI Helium SSME Post Operations Purge Regulator

⑭
O_2 / H_2 Propellant Flow Control Valve

Dual solenoid water control valves are used in the environmental control system. During re-entry, Eaton wing vent relief doors are employed. From lift-off to landing, Eaton components play a vital role in each shuttle mission.

The diversity of the El Segundo facility product lines is extensive. The company supplies pressure regulators, thruster control valves, isolation valves and ordnance devices for communications, satellites, missiles and expendable launch vehicles. Eaton is a leading supplier of aircraft fuel components, in-flight refueling receptacles, bleed air valves, door snubbers, cargo rollers and oxygen door latches for military and commercial aircraft. The company's aerospace valve technology has been successfully applied to the air-bleed control requirements of the U.S. Navy gas turbine powered ships, the FFG-7 frigates, DD963 destroyers and CG47 missile cruisers. Eaton is the exclusive supplier of 925 degree F pressure regulators, relief valves and control valves for the Prairie and Masker systems.

In 1986, Eaton acquired the Space Products line from Hydraulic Research. This provided Eaton with sophisticated, small size, torque motor and solenoid operated valves used on numerous satellite and space vehicles, including the space shuttle, the Manned Maneuvering Unit, TIROS, MMS, GRO, COBE, HBO, ANIK-E, Intelsat, GPS, Cassini, Peace Courage and many more. Lower flowrate, low-leakage product line capability augments the existing wide variety of larger, higher flow capability space qualified products.

Eaton is currently the leader in the development of the latest state-of-the-art flow and pressure control technology with its NASA-funded technology development program for the Electronically Variable Pressure Regulator (EVPR). All desired control functions are provided by the same basic device simply by using the appropriate sensor, e.g. pressure transducer, flow sensor or temperature probe as the closed-loop feedback control element to the micro-controller device. The potential for using common mechanical assemblies with custom computer control packages to provide precise pressure, flow and temperature regulation, or as a shut-off device, is believed to be the wave of the future in components technology for space, marine and energy-related applications.

The Pressure Sensor facility of Eaton is located in Bethel, Connecticut. The Bethel facility produces sensors for use in severe environments - those encountered in space, missiles, aircraft, aerospace engines and naval underwater applications.

Eaton has been providing space qualified sensors since the Apollo launch and now provides a total of 129-qualified sensors for shuttle-related applications. The space shuttle utilizes eight control pressure conditioners to regulate raw shuttle power and provide a very precise 24 volt supply to the 16 control pressure transducers. The control pressure conditioners utilize the control pressure transducer signal outputs to amplify and integrate them into two analog data channels. These discrete signals are used to control heaters, internal to the cryogenic tanks, which maintain tank pressure within prescribed boundaries to feed the fuel cells that generate the main power requirements during the entire flight.

Pressure transducers, 44 each flight, are used in the primary and vernier thrusters of the shuttle. The result of the chamber pressure monitoring by the transducers is used to indicate to system logic that the thruster has fired for the prescribed length of time.

A total of 60 dual output SSME pressure transducers are used throughout the three main engines to monitor and control a number of critical functions. Finally, a single output version of the SSME pressure transducer designed to operate at -160 degrees F is used for monitoring hydrogen fuel pressure being fed to the main engines.

Eaton also produces special sensors for the Trident, Peacekeeper, Minuteman and the Standard Missile, in addition to providing sensors for naval shipboard use, both surface and underwater applications, including guided missile cruisers, nuclear submarines and aircraft carriers.

Aircraft engines are a special challenge for the sensor manufacturer and Eaton is proud to be a major supplier of sensors to a number of the world's leading makers of aircraft engines and APU's. Other applications in severe environments, such as the following for commercial and military aircraft are: monitoring hydraulic pressure in a wide variety of applications, lubricating oil pressure and differential pressure across valves and filters; generating warning signals for alarm conditions such as high or low pressure. In addition, there are applications for fuel systems which include fuel pressure, system differential pressure and fuel level, as well as position sensors to detect valve spool and throttle position. Options include multiple redundant outputs, flame proofing and radiation hardening. Some of the aircraft involved are Boeing's 737, 747, 757, 767 and 777 aircraft; General Dynamics F-16; Northrop's F/A18 and the B-1 and B-2 Bombers.

Both Eaton's El Segundo and Bethel facilities, as suppliers to the NASA subcontractors since the beginning of the space shuttle program, are fully conversant with the NASA quality and traceability requirements. In anticipation of continuing in this important role, Eaton maintains a team of managers who properly monitor the various programs, manufacturing and inspection functions to assure the timely delivery of a reliable, quality product. Eaton fully understands the NASA space environment and its unique requirements, and looks forward to being part of the NASA team of the future.

Index

Sponsoring Companies

Aerotherm Corporation, 140
AlliedSignal Aerospace, 136
Applied Solar Energy Corp., 156
Barrios Technology, 163
Barry Controls, 150
Booz.Allen & Hamilton, 144
DATATAPE Incorporated, 132
Digital, 159
Eaton Corporation, 165
Evans & Sutherland, 148
Hughes Electronics Corp., 114
Harris Corporation, 152
Instron Corporation, 110
Johnson Engineering Corp., 134
Link, 160
ManTech International Corporation, 142
Martin Marietta Corporation, 118
McDonnell Douglas Aerospace, 154
Motorola, 122
Network Systems Corp. 126
Rockwell, 106
Silicon Graphics, Inc., 164
SPACEHAB, 146
Space Vector Corporation, 138
USBI Co., 162

Space History Photographs

Agena, 29, 30
Aldrin, Edwin, 31, 46-51
Allen, Joseph, 77
Anders, William A., 38
Apollo, 32,35,37,40, 41, 42,52
Apollo 13, 55, 56
Apollo 15 Scientific Module, 61
Apollo-Soyuz, 69-72
Armstrong, Neil, 29, 43, 47, 50
Atlantis, 81
Bassett, Charles, 30
Bean, Alan L., 52, 53
Bluford, Guion, 78
Bobrov, Yevgenly, 70
Borman, Frank, 28,38
Brand, Vance D., 69, 77
Brooks, Jack, 95
Buchli, James, 81
"Bumper" rocket, 2
Bush, George, 88
Carpenter, Scott, 20
Cernan, Eugene A., 30, 41
Chaffee, Rober B. 33
Challenger, 78, 83, 85, 86, 87
Clinton, Bill, 95
Cohen, Aaron, 1
Collins, Michael, 28, 31, 43, 50
Columbia, shuttle, 73, 74, 75
Comet Kohoutek, 68
Conrad, Charles, 27, 31, 52, 66
Cooper, Gordon, 21, 27
Crippen, Robert L., 75
Cunningham, Walter, 37
Dezhurov, Vladimir, 95
Discovery, 88, 96
Duke, Charles, 61
Dunbar, Bonnie, 80, 95
Eagle, 45, 48
Earth, 39, 62
Echo 1, 11
Eisele, Don F., 37
Elms, James, 29, 31
Enos, 18
Enterprise shuttle, 73
Explorer 1, 6-7
Footprint on moon, viii
Fullerton, Gordon, 73
Furrer, Reinhard, 80
Galileo, 90
Gamma Ray Observatory, 90
Garn, Jake, 80
Garriott, Owen K., 67
Gemini-Titan 3, 25
Glenn, John, 19, 20
Gilruth, Robert, 9, 21, 31, 46, 50
Goldin, Daniel, 95
Gordon, Richard, 31, 52
Griffin, Gerald D., 57
Grissom, Gus, 17, 33
Haise, Fred W., 54, 57, 73
Ham, 11
Hartsfield, Henry, 76, 80
Hauchk, Frederick, 77
Hoffman, Jeffrey, 91
Holmes, Brainerd, 9
Hubble Space Telescope, 91
Irwin, James, 60
Jarvis, Gregory, 82, 84
Johnson, Lyndon B., 21, 27
Johnson Space Center, 10
Kennedy, John F., 15, 16, 21
Kerwin, Joseph P., 66
Kraft, Chris, 22, 29, 51
Kranz, Eugene A., 57
Krikalev, Sergei, 92, 95
Kubasov, Valeri N., 69
Langley Research Center, 9
Lenoir, William, 77
Leonov, Aleksei, 69, 70, 72
Lovell, James, 28, 31, 38, 54, 57
Lunar Rover, 60, 61
Lunney, Glynn S., 57
M15 star cluster, 105
M100 galaxy, 92
Magellan, 89
Manned Maneuvering Unit, 78
Mars, Future Exploration, 103-105
Matthews, Charles W., 31
Mattingly, Thomas, 76, 81
McAuliffe, Sharon Christa, 82, 84
McCandless, Bruce, 78
McDivitt, James A., 26, 39
McNair, Ronald E., 82, 84
Mercury, 105
Mercury 7 Astronauts (Group), 10
Mercury Redstone, unmanned, 13
Mercury Redstone 3, 14
Mercury-Atlas 6, 19
Milky Way Galaxy, 11
Mir, 93, 96
Mission Control, 38, 42, 46, 79
Mitchell, Edgar, 58
Moon, 36, 49, 54, 63, 98
Moon, Future Exploration, 99-102
Morgan, Barbara, 82
Musgrave, Story, 77, 91
NACA committee, (1934), 8
Nagel, Steven R., 80
Nelson, George, 79
Nicollier, Claude, 91
Nixon, Richard M., 50, 57
Ockels, Nubbo J., 80
O-ring joint, 87
Onizuka, Ellison, 81, 82, 84
Overmyer, Robert, 77
Paine, Thomas O., 57
Payton, Gary, 81
Plaque, Moon, 47, 60
Powers, John (Shorty), 9
Reagan, Ronald, 76
Reeder, Lloyd, 50
Resnik, Judith A., 82, 84
Ride, Sally, 77
Robot arm, shuttle, 76
Rocks, moon, 58, 63
Ross, Jerry, 81
Salyut 1, 59
Satellites, 23, 80, 88
Saturn, 105
Saturn rocket, 32, 35, 51, 52
Schirra, Walter M., 21, 22, 28, 37
Schlegel, Hans, 92
Schmitt, Harrison, 62
Schweickart, Russell L., 41
Scobee, Francis R. "Dick", 82, 84
Scott, David R., 29, 41
Seddon, Rhea, 80
See, Elliott, 30
Sega, Ronald, 92
Shepard, Alan, 12, 15, 58
Shriver, Loren, 81
Sjoberg, Sigurd A., 57
Skylab, 64-68
Slayton, Donald K., 50, 69, 72
Smith, Michael J., 82, 84
Solar Array, 79
Solar Corona, 45
Solar Eruption, 66
Solovyov, Anatoly, 95
Splashdown, Apollo, 56
Spacelab, 80
Space Station, 94, 96, 97
Sputnik, 4
Stafford, Thomas, 28, 30, 41, 69, 70, 72
Strekalov, Gennady, 95
Surveyor, 35
Swigert, John L., 54, 57
Syromyatnikov, Vladimir, 70
Taurus-Littrow, 62
Thagard, Norman, 77, 95
Thornton, Kathryn, 91
UARS, 90
Ulysses, 90
van Hoften, James, 79
Venus, 89, 105
von Braun, Wernher, 3
Webb, James E., 14, 22
White, Edward H., 24, 26, 27, 33
White, Robert, 70
Williams, Walter C., 9, 22
Windler, Milton L., 57
Young, John, 31, 41, 75